T0137072

Information Theory and Selected Applications

Arieh Ben-Naim

Information Theory and Selected Applications

Arieh Ben-Naim
Department of Physical Chemistry
The Hebrew University of Jerusalem
Jerusalem, Israel

ISBN 978-3-031-21278-9 ISBN 978-3-031-21276-5 (eBook)
https://doi.org/10.1007/978-3-031-21276-5

This Springer imprint is published by the registered company Springer Nature Switzerland AG
The registered company address is: Gewerbestrasse 11, 6330 Cham, Switzerland

אִם אֵין בִּינָה, אֵין דָּעַת. אִם אֵין דַּעַת, אֵין בִּינָה.

פרקי אבות, פרק ג, משנה יז

If there is no knowledge there is no understanding;
If there is no understanding there is no knowledge.

This book is dedicated to all those who want to know how to apply Information Theory correctly, avoiding confusing "Information" with "Shannon's Measure of Information" and Shannon's Measure of Information with Entropy

Preface

This book is a sequel to my previous book: *Information Theory, Part I: An Introduction to the Fundamental Concepts,* published in 2017. It consists of six different applications of Information Theory (IT). Chapter 1 is an introductory chapter. As in Part I, I have written this chapter to avoid the confusion that exists in the literature about the concepts of "Information," Shannon's Measure of Information (SMI), and the (thermodynamic) entropy. Although these concepts are related to each other, they are quite distinct. Information is the most general concept which includes all types of information. The SMI is not *information,* but a measure of a specific kind of information associated with any possible probability distribution. Similarly, SMI is not entropy! Entropy is a specific measure of information defined on a specific set of probability distributions. Chapter 1 aims to clarify these three concepts, as well as the unit of information known as the bit. It also introduces the concept of frustration and discusses the question of the quantification of this concept within IT. Chapter 2 focuses on the interpretation of the entropy of systems of interacting particles in terms of the SMI and of mutual information.

Chapter 3 is a study of a small group of spins at some specific configurations. It examines the question of the possibility of measuring the extent of frustration using mutual information.

Chapter 4 discusses some classical examples of processes of mixing and assimilation for which the entropy changes are interpreted in terms of SMI.

Chapter 5 describes a few binding systems and the interpretation of cooperativity phenomena in terms of mutual information.

Chapter 6 discusses the general method of using maximum SMI in order to find the "best-guess" probability distribution.

The book may be read by any scientist who is interested in IT and in its potential applications. It might also be useful to those who are considering the application of IT in their own fields of interest.

The reader is urged to read and absorb Chap. 1 first. After doing so, the reader can choose to read any other chapter of the book in whatever order of his/her choosing.

I also urge the reader who is familiar with an interesting application of IT to write to me. I might include these additional applications in a future edition of this book, or an entirely different book.

Jerusalem, Israel

Arieh Ben-Naim
https://www.ariehbennaim.com

Acknowledgments I am grateful to many who read parts or all the manuscript and offered important comments. In particular, I am grateful to: Diego Casadei, Claude Dufour, David Gmach, Jose Angel Sordo, and Thierry Lorho.

As always, I am very grateful for the graceful help from my wife, Ruby, and for her unwavering involvement in every stage of the writing, typing, editing, re-editing, and polishing of the book.

Contents

Abbreviations

1D	One dimensional
20Q	Twenty questions
GPF	Grand partition function
HR	Hard rod
ID	Indistinguishable
Ig	Ideal gas
IT	Information Theory
MI	Mutual Information
MSD	Micelle-size-distribution
nn	Nearest neighbors
nnn	Next nearest neighbors
PF	Partition function
rv	Random variable
SMI	Shannon's Measure of Information

Chapter 1
Introduction and Caveats

1.1 A Bit of Information About the Bit in Information Theory and the Binary Digit

The word "bit" in English has several meanings. Normally, it expresses a small amount of something. In Information Theory (IT), the "bit" has two different definitions; one as a "*bi*nary-dig*it*," and the second as a *unit of information.* Unfortunately, these two distinct definitions are sometimes confused. In this section we discuss these two meanings of the "*bit*". See also Ben-Naim [1].

(i) ***The bit in communication***

Originally, a *bit* was an abbreviation for "*bi*nary un*it*." In a decimal number we use 10 digits, or 10 symbols: 0, 1, 2, 3, 4, 5, 6, 7, 8, 9. When we write a decimal number (base 10) such as 25, we mean $25 = 5 + 20 = 5 \times 10^0 + 2 \times 10^1$, and similarly $256 = 6 + 50 + 200 = 6 \times 10^0 + 5 \times 10^1 + 2 \times 10^2$. In binary numbers we express a number with only two digits; 0 and 1. In this case we use the powers of 2, instead of powers of ten.

Examples of numbers in the two bases are

Decimal number	Binary number
0	0
1	1
2	$10 = 0 + 1 \times 2^1$ $(= 2)$
3	$11 = 1 + 1 \times 2^1 (= 3)$
4	$100 = 0 + 0 \times 2^1 + 1 \times 2^2$ $(= 4)$
5	$101 = 1 + 0 \times 2^1 + 1 \times 2^2$ $(= 5)$

(continued)

A. Ben-Naim, *Information Theory and Selected Applications*,
https://doi.org/10.1007/978-3-031-21276-5_1

(continued)

Decimal number	Binary number
$\vdots$	
10	$1010 = 0 + 1 \times 2^1 + 0 \times 2^2 + 1 \times 2^3 (= 10)$
15	$1111 = 1 + 1 \times 2^1 + 1 \times 2^2 + 1 \times 2^3 (= 15)$

In computation and telecommunication, we use the binary language which essentially is a two-symbol or two-letter language. These could be "Yes" and "No, "0" and "1," or a magnet "up" and "down," etc. When we communicate a message in English we first translate each letter into a *code*, and transmit the encoded message to the receiving terminal, where it is decoded back into English letters.

In the process of transmission of information, we are usually interested in achieving a highest accuracy (or faithfulness) of transmission of a given information (in spite of noise), at a lowest cost.

One such code used in telegraph is the Morse code, Table 1.1, which uses a sequence of dots and dashes as code-word for representing the different letters. Clearly, the "length" of a dot, is shorter than that of a dash, let's say a dot takes a unit of time to transmit, then the dash will take three units. Clearly, to transmit a given text we would like to have a shorter code-word for the more frequent letters, and longer code-word for the less frequent letters. This is roughly how the Morse code was constructed. As you can see from the table, this proportionality is not always true in the Morse code.

Another code known as ASCII (American Standard Code for Information Interchange) is used in electronic communication.

Each of the digit, 0 or 1 is referred to as a *bit*. Thus, when we send a sequence of bits such as: 1, 0, 1, 1, 0, 1, 0, 1, 1, 0, we say that we sent 10 bits. This is equivalent to saying that we sent ten symbols, which happen to be zeros and ones. Sometimes, it is said that we sent ten bits of *information*. This is true only when we use a binary digit as a "unit" of information. Another meaning of the "bit," as a unit of information, follows.

(ii) ***The bit as a unit of information***

We begin with the statement that the "*bit*" in IT is a *measure of information*, and it is not the same as the "bit" in "*bi*nary dig*it*."

The definition of the *bit* in IT arose from Shannon's measure of information (SMI) when applied to the case of two outcomes, Shannon [2]. If an experiment, or a random variable (rv), has only two possible outcomes, say, 1 and 2, with probabilities p_1 and p_2, the corresponding SMI is:

$$H(p_1, p_2) = -p_1 \log p_2 - p_2 \log p_2 \tag{1.1}$$

Since $p_1 = 1 - p_2$ we have in fact, a one-parameter function. Setting $p_1 = p$, and $p_2 = 1 - p$, we rewrite (1.1) as:

Table 1.1 Frequencies and Morse code for English letters

Letter	Frequency (%)	Morse code
E	12.60	•
T	9.37	—
A	8.34	•—
O	7.70	———
N	6.80	—•
I	6.71	••
H	6.11	••••
S	6.11	•••
R	5.68	•—•
L	4.24	•—••
D	4.14	—••
U	2.85	••—
C	2.73	—•—•
M	2.53	——
W	2.34	•——
Y	2.04	—•——
F	2.03	••—•
G	1.92	——•
P	1.66	•——•
B	1.54	—•••
V	1.06	•••—
K	0.87	—•—
J	0.23	•———
X	0.20	—••—
Q	0.09	——•—
Z	0.06	——••

$$H(p) = -p \log p - (1 - p) \log(1 - p) \tag{1.2}$$

Here, log is the logarithm with base 2. The function $H(p)$ is shown in Fig. 1.1. Note that this function has a single maximum at $p_{\max} = \frac{1}{2}$, and the corresponding value of $H(p)$ is: $H(p_{\max}) = 1$ (when base 2 is used).

The interpretation of $H\left(p_{\max} = \frac{1}{2}\right) = 1$ as a unit of information follows from the interpretation of the SMI in the general case of n outcomes as the amount of information associated with a probability distribution. In the case of the two outcomes the unit of information is called the *bit*. This is the amount of information one gets when asking a binary question about two outcomes having *equal probabilities*.

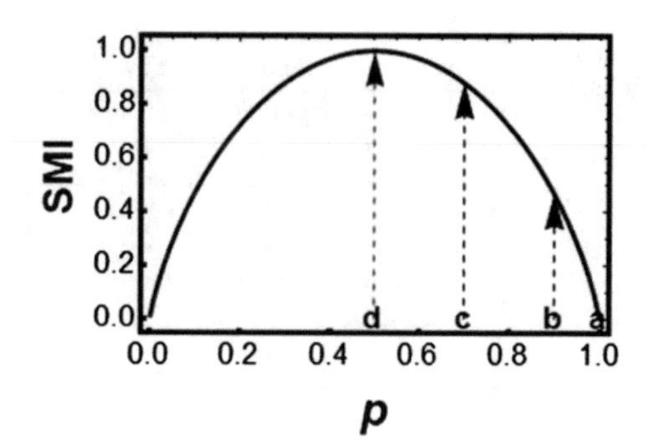

Fig. 1.1 The function $H(p)$, Eq. (1.2) and the values of the SMI for the coins a, b, c, d

The emphasis on "equal probabilities" is important. It is very common to find statements, in popular science books referring to a "bit" as the amount of information one gets to a binary question. To see why this is not true consider the following examples of coins with different probability distributions.

Suppose we have four coins with distributions given in Table 1.2.

Now, suppose you play the following game ten times: You are given the distribution as in Table 1.2 and you have to guess the outcome of throwing the coin. You can ask either "is the outcome H?" or "is the outcome T?" And you can get either a Yes, or No answer.

To make the game more dramatic let us add that you get a dollar when you get an answer Yes, and get nothing when the answer is No. Clearly, your interest is to ask questions such that you will earn maximum dollars.

Coin a: In this case, you know with certainty that the coin will always show H. Therefore, you will always ask, "is the outcome H?" If you play this game ten times, you will get ten answers of Yes.

Yes, Yes, Yes, Yes, Yes, Yes, Yes, Yes, Yes, Yes.

In terms of dollars, you get one dollar each time you play the game, hence, you can translate the sequence of Yes and No into a sequence of binary digits, in this case:

1, 1, 1, 1, 1, 1, 1, 1, 1, 1

Why did you get a dollar each time? Because by *knowing* the distribution (1, 0) you got a lot of information on the outcomes (or equivalent, your lack of information on the outcomes was minimal). You used this information to maximize your earnings.

Coin b: In this case, you will also ask: "is the outcomes H?" Here, you are not guaranteed that you will always get a Yes answer. If you play ten times it is likely that

Table 1.2 Four coins and the probability distribution of their outcomes

Coin	Probability distribution
Coin a	$p_H = 1,\ p_T = 0$
Coin b	$p_H = 0.9,\ p_T = 0.1$
Coin c	$p_H = 0.7,\ p_T = 0.3$
Coin d	$p_H = 0.5,\ p_T = 0.5$

you will get on average about one No answer, and nine Yes answers. The sequence of answers you get in terms of dollars will look like:

1, 1, 1, 0, 1, 1, 1, 1, 1, 1

So you did quite well by playing this game. You won most of the time. Why? Because you knew the distribution which gave you a lot of information. It is not as good as the amount of information you had in the case of **coin a**, but it is still quite a lot.

Coin c: In this case you will also be wise to ask each time: "is the outcome H?" The series of answers you will get when translated from (Yes, No) into (1, 0) will, most likely look like:

1, 1, 1, 0, 1, 0, 1, 0, 1, 1

A simple calculation shows that, in this case, you won on average seven dollars in playing ten games. Here, the amount of information you got, by knowing the distribution is less than in the case of **coin b**, but better than the following case.

Coin d: Clearly, in this case, whatever your questions will be: "Is the outcome H" or "is the outcome T?" You will get, on average about five Yes answers. In terms of dollars, you obtain a sequence of binary digits which looks like:

1, 0, 1, 1, 0, 1, 0, 0, 1, 0

Here, you get on average about five Yes, and five No answers. In this case, you earn the least amount of dollars. (In a more realistic game, you get a dollar for a Yes answer, and you pay a dollar for a No answer. In such a game you get on average nothing by playing this game).

The reason you earn minimal amount of dollars is due to scant amount of information you obtained by knowing the distribution. You can also say that the amount of information you *lack* is maximal in this case.

Thus, we see that by playing with coins a, b, c, d, you *have* maximum information in a, and minimum information in d. Equivalently, the *lack* of information is minimum in case a, and maximum in case d. The different amounts of information are shown by the different arrows in Fig. 1.1 (with $p = p_H$).

The amount of information you lacked is measured by the function H. The minimum is zero and the maximum is one *bit*. The *bit* is the amount of information you obtained by an answer to a binary question about two equally probable outcomes.

Thus, in the case of **coin d** you obtain one bit for each question you ask (or the amount of information you lacked in each game is one bit). In playing the same game 10 times, you obtain *ten bits* of information.

Take a look at the sequences of ones and zeros that we got by playing with the four different coins. In each of these games, we played ten times, and the number of "*binary digits*" we got was always *ten*. The *amount of information* we got by playing with the four coins was different. The SMI values for these four coins are:

$$H(a) = 0 \text{ bits},$$
$$H(b) = 0.47 \text{ bits},$$
$$H(c) = 0.88 \text{ bits},$$

$$H(d) = 1 \text{ bit} \tag{1.3}$$

The value of H in this case is the amount of information you obtain from a binary question. It is zero for **coin a** (since you already knew the result), and it is one for **coin d**. Thus, it is only in the case of **coin d** that the number of *bits*, in terms of *binary digits* (ten), is equal to the number of *bits* in the informational theoretical sense.

We have discussed only the case of two outcomes; this is the case for which the bit is defined. However, for any experiment (or a game or a random variable) we obtain some information about the outcomes when we have the distribution. The best way to understand this amount of information is with a 20-question (20Q) game. One person chooses one out of n outcomes, and the second person has to ask binary questions in order to find out which outcome was chosen. Clearly, the more information you lack the more questions you must ask in order to obtain the lacking information. It turns out that if you ask *smart questions*, you get the maximum of one bit per question. Therefore, by asking smart questions you will need the fewest questions in order to find out the lacking information. This number is the SMI for the general case of n outcomes. For details, see Ben-Naim [1], Chap. 2, and Ben-Naim [3].

1.2 Misinterpretation of Probability as SMI and SMI as Probability

There are the two, very common mistakes; the first is to interpret the SMI as an *average information*, and the second is to interpret the SMI as probability. We will discuss each of these mistakes separately.

(i) **p_i is not a measure of information, and $-\log p_i$ is not measured in bits**

In numerous textbooks on IT, as well as in popular science books one can find a description of $-\log p_i$ as a measure of information associated with the event i, hence, the SMI $= -\sum p_i \log p_i$ is interpreted as an *average* information. This erroneous misinterpretation of SMI is discussed further in Ben-Naim [1]. Here, we focus only on the single term $-\log p_i$, which is sometimes referred to as "self-information," or the amount of information you get when you know that the event i occurs. Some even assign to this the term a value in units of bits.

Here is how "self-information" is introduced in Wikipedia:

> Definition: Claude Shannon's definition of self-information was chosen to meet several axioms:
>
> If two independent events are measured separately, the total amount of information is the sum of the self-information of the individual events…given an event x with probability P, the information content is defined as follows:
>
> $$I_X(x) = -\log(P_X(x))$$

This whole quotation is not only untrue; it is misleading as well. First of all, Shannon *never defined self-information*, (neither in the original article, Shannon [2], nor in Shannon and Weaver [4], and, of course, this was never chosen to meet "several axioms."

Shannon searched for a *measure of information* based on the whole distribution and not for a single event. His conditions (as in Shannon [2]: "*it is reasonable to require of it the following properties*"), were entirely different from the conditions or requirements stated in abovementioned quotation.

If an event with a probability 1 occurs, it is not surprising, it is very much expected, but it is not true that it yields *no information.* When I hear that an event x with probability 100% occurred, I *obtained* the information that "x occurred".

If an event with lower probability occurred, I am more *surprised.* This it is true. But it is not true that I obtained *more information*!

Suppose that we have four dice with different probability distributions, say

$$\begin{aligned}
&\text{die A: } p_1 = 1,\ p_2 = p_3 = p_4 = p_5 = p_6 = 0\\
&\text{die B: } p_1 = 0.9,\ p_2 = 0.1,\ p_3 = p_4 = p_5 = p_6 = 0\\
&\text{die C: } p_1 = 0.8,\ p_2 = 0.2,\ p_3 = p_4 = p_5 = p_6 = 0\\
&\text{die D: } p_1 = 0.7,\ p_2 = 0.3,\ p_3 = p_4 = p_5 = p_6 = 0
\end{aligned}$$

When I tell you that the outcome "1" had occurred, the *information* you obtained was: the outcome "1" occurred. This *information* is independent of the probability of the *outcome* "1." Also, the *amount* of *information* you got is independent of the probability. When I tell you that the outcome "2" had occurred, the information you obtained was: "the outcome "2" occurred," this information is independent of the probability of the outcome "2." Also, the *amount* of *information* is independent of the probability. Looking at the *entire distributions* of the different dice, Shannon's Measure of Information (SMI) increases from A to B, and from B to C, and so on. This *measure of information* depends on the *entire distribution,* and it gets a maximum value when the distribution is uniform, i.e. $p_i = \frac{1}{6}$ for $i = 1, 2, \ldots, 6$.

When p_1 decreases from 0.8 to 0.7, the SMI of the *entire distribution increases,* not because the probability of the outcome "1" decreased, but because the entire distribution changed towards more uniformity (in going from die C to D).

Similarly, when p_2 decreases form 0.3 to 0.2, the SMI of the *entire distribution decreases* because the distribution became less uniform in going from die D to C.

The most important lesson one learns from this example is that SMI is *defined* on the *entire distribution,* not on the probability of a single event. This is exactly what Shannon sought and found, a measure of information associated with the entire probability distribution.

It is often said that getting information about the outcomes removes the uncertainty. This is true provided we specify uncertainty, but with respect to what?

Suppose that I informed you that the outcome "1" had occurred when I threw each of the dice above, which uncertainty was removed?

1. The uncertainty about the color of the die?
2. The uncertainty about outcome "1?"

3. The uncertainty about the outcome "6?"
4. The average uncertainty about all the outcomes.

Obviously, if I was not told about the color of the dice, the uncertainty about the color was not removed. All the three other uncertainties were removed. However, the *amount* of uncertainty removed is the same about the outcome "1" for all the dice, and the amount of information I received by being informed on the specific outcome that occurred (any of the outcomes), is the same and is equal to the SMI of the specific die (not a specific outcome).

Here is a typical statement about the relationship between probability and information. In Rosenhouse's book [16], *The Monty Hall Problem*, page 104 we find:

> Intuitively there ought to be an inverse correlation between probability and information. That is, you receive more information from learning that a low-probability event has occurred than you do from learning that a high-probability event has occurred. When it is nearly certain that X will occur, you do not learn very much upon hearing that X has in fact occurred. If you hear instead that Y, a highly unlikely alternative to X, has occurred, then you have learned something far more informative.

Obviously, the author confuses here the "extent of surprise" one gets from learning that an event has occurred, with the "amount of information" one gets. It is strange that the above statement appears in a book on the Monty Hall problem, which is a perfect example where one can clearly understand the difference between information, probability and the amount of information. We shall discuss the Monty Hall problem Appendix 2.

How was this error born? I believe the answer to this question is as follows. Suppose we have an experiment with n equally probable outcomes. In this case, the probability of each outcome is:

$$p_i = \frac{1}{n} \text{ for each } i = 1, 2, \ldots, n \tag{1.4}$$

The SMI for such an experiment is:

$$\text{SMI} = -\sum_{i=1}^{n} p_i \log p_i = -\sum_{i=1}^{n} \frac{1}{n} \log \frac{1}{n} = \log n \tag{1.5}$$

Thus, for n equally probable outcomes the SMI is $\log n$. Now, since all the probabilities p_i are equal, Eq. (1.4), we can denote $p_0 = \frac{1}{n}$, and rewrite (1.5) as:

$$\text{SMI} = \log n = \log \frac{1}{p_0} = -\log p_0 \tag{1.6}$$

which is also correct. p_0 is indeed the probability of a single event (any event), and it is related to the SMI, and therefore this SMI is measured in units of bits. It is clear that p_0 is related to SMI not because it is a "probability of a single event," but because it represents the *entire distribution.*

From Eq. (1.6) it is easy to conclude (erroneously) that $-\log p_i$ for any event in any experiment is also an SMI and should be measured in bits. This conclusion is incorrect for an experiment for which all the events are not equally probable. Thus, in general, $-\log p_i$ is not "*self-information.*"

Mackay [5] refers to $-\log p_i$ as "Shannon information content of an outcome," and even gives a table of such values for the letters in English. I believe this term is potentially misleading.

Some use the uncertainty interpretation of SMI (see Sect. 1.2 of Ben-Naim [1]) to justify the interpretation of $-\log p_i$ as the amount of information. The argument is as follows: If you are told that the event *i* occurred, then the uncertainty about that event was removed. Indeed, the uncertainty about the event *i*, is measured by the quantity $-\log p_i$. However, according to the uncertainty-interpretation of the SMI, the *uncertainty removed* is about all the results of that experiment. The correct interpretation of SMI is not just "uncertainty" but the average uncertainty about the occurrence of all the possible outcomes. This comment applies to the uncertainty-interpretation of entropy.

When I am told that: "Result *i* occurred," the *information* I got is simply that result *i* occurred. The *meaning* of this information is different for different events, but the *amount* of information is the same, independent of *i*.

Here is an example from Stone's book [17]. Consider an unfair (or a biased) coin with probabilities $p_H = 0.9$ and $p_T = 0.1$. Stone tells us that: If the outcome is a head then the amount of Shannon's information is:

$$-\log 0.9 = 0.15\,\text{bits per head}$$

If the outcome is a tail, then the amount of Shannon information is:

$$-\log 0.1 = 3.32\,\text{bits per tail}$$

Then, the author adds the comment:

> Notice that more information is associated with the more surprising outcome (a tail, in this case).

This statement clearly confuses the *informational* interpretation, and the *surprisal* interpretation of SMI. The surprisal (as uncertainty and unlikelihood), applies to each individual event. The SMI is an *average* of all the surprisals, uncertainties, or unlikelihood of all the events. Therefore, the SMI is a property of the entire probability distribution. The informational interpretation also applies to the entire distribution and not to the individual events. See Ben-Naim [1], Sect. 1.2.

Thus, the two "estimates" of the "Shannon information" of the two outcomes cannot be measured in terms of *bits*—in the sense units of information! The SMI associates with this particular coin was already calculated above for coin b; SMI = 0.47 bits. This quantity is defined on the entire distribution, not on each single event.

Perhaps, one can say that since $-\log 0.1 = 3.32$ is a bigger number than $-\log 0.9 = 0.15$, the expression for the first would require bigger binary-number, i.e., more binary digits than for the smaller number. This reasoning however amounts to confusing the bits in IT with the bit as binary digit.

(ii) **SMI is not a probability**

In the beginning of this section we claimed that probability in general, may not be interpreted as SMI. It is true that in a special case when all $p_i = p_0 = \frac{1}{n}$, then $-\log p_0$ may be interpreted as SMI. However, in general $-\log p_i$ is not SMI. From this particular example, one cannot conclude that SMI is, in general, probability.

The association of SMI with probability is probably due to Brillouin [6]. On page 120 of his book "Science and Information Theory," we find:

> The probability has a natural tendency to increase, and so does entropy. The exact relation is given by the famous Boltzmann-Planck formula:
>
> $$S = k \ln P \tag{1.7}$$

It is difficult to overestimate the amount of misinformation that is packed in these two sentences. Probability has no natural tendency to increase! Probability does not behave as entropy! There is no exact relationship between entropy and probability! The quoted formula is not the Boltzmann-Planck formula.

The correct Boltzmann-Planck relationship for the entropy is $S = k \ln W$, where W is the total number of accessible microstates in the system. This relationship is a special case SMI for the case when all the events have equal probabilities. As we showed above, in general, probability is not SMI (except when $p_i = p_0 = \frac{1}{n}$).

Here, we claim that entropy (being a special case of SMI) is never related to probability by an equation $S = k \ln P$.

The simplest reason for my claim is that probability is a positive number between 0 to 1. Therefore, $\ln P$ varies between minus infinity to 0. Entropy, as well as SMI is always a positive number greater or equal to 0. More on this in Ben-Naim [7].

1.3 SMI, in General Is not Entropy. Entropy Is a Special Case of SMI

Shannon defined a quantity H as a function of the entire distribution

$$H(p_1, \ldots, p_n) = -\sum p_i \log p_i \tag{1.8}$$

It is clear that for any given distribution $(p_1, \ldots, p_n)$ one can define the corresponding SMI. The definition of the function $H(p_1, \ldots, p_n)$ was also generalized to the case of a continuous random variable. In this case, the SMI is a functional defined for any distribution density $f(x)$:

$$H[f(x)] = -\int f(x)\log f(x)dx \tag{1.9}$$

Clearly, this functional is not always a positive number and it is not always a finite quantity. Some of the mathematical problems in the definition of the SMI for the continuous case were discussed in Chap. 2 and Appendix C of Ben-Naim [1]. We shall sometimes use the notation $H(p_1, \ldots, p_n)$, sometimes we use the notation either $H(X)$ or SMI (X), for the SMI of the random variable X.

Entropy is defined on a very special set of probability distributions. For classical systems of ideal gases, the relevant distributions are of the locations and momenta of all particles of the system at equilibrium. For such distributions the entropy is proportional to the corresponding SMI (see Chap. 5 of Ben-Naim [1]).

Thus, one can talk on the SMI of a die, or a coin or any other experiment. All these are not entropies.

It is unfortunate that Shannon himself called his measure of information, entropy. This was a great mistake which caused great confusion in both thermodynamics and in IT [examples of such confusion are discussed in Ben-Naim [7–12]].

In this book we discuss both SMI and entropy. Whenever we discuss entropy we mean a special case of SMI. Whenever we discuss SMI, we mean SMI, not entropy.

There are many formulas which look like entropy and are called entropy, but they are not entropy. Examples are the Tsallis entropy, Rény entropy, and others. They are not entropy in the sense that they are not equivalent to Clausius's entropy, or the entropy, defined as the SMI based on the distribution of locations and momenta of all particles at equilibrium.

To see why calling any SMI entropy might be confusing, consider a fair die, i.e. all outcomes have the same probability. If one calls SMI "entropy," then "ln 6" is called the entropy of a fair die which is very different from the thermodynamic entropy of the die. To demonstrate why this identification might lead to awkward statements, consider the following "processes."

1. Suppose that the temperature of the die increases, does the entropy of the die change?
2. Suppose the die is made of clay and it breaks into two parts when it falls on the ground, will the entropy of the die change in this process?
3. Suppose the die is distorted in such a way that its total volume and surface are nearly unchanged. However, the probability distribution of the six outcomes might change considerably. Does the entropy of the die change?

Clearly, one cannot answer any of these questions without making a distinction between the SMI associated with the distribution of the appearance of the number of dots on the faces of the die, and the entropy which is defined by the distribution of locations and momenta. The SMI of the die changes in processes 2 and 3, but not in 1. The entropy of the die changes in process 1, but it is almost unchanged in 2 and 3.

Similar confusion had occurred in the Boltzmann H-Theorem which is discussed in Ben-Naim [7, 13].

1.4 The "Vennity" of Using Venn Diagrams in Representing Dependence Between Random Variables

Venn diagrams are useful for representing various operations between sets or events. In Fig. 1.2, we show a few examples of Venn diagram for the operations:

(a) union (or sum) of two events $A \cup B$
(b) intersection (or product) of two events $A \cap B$
(c) disjoint (or mutually exclusive) of two events $A \cap B = \phi$
(d) inclusion; B is contained in A, $B \subset A$.

In using the Venn diagram, a region represents an event and the area of the region is proportional to the probability of that event. For instance; we throw a dart on a board, and we know that the dart hit the board. This means that the certainty event Ω, which is the area of the entire board, is assigned the probability one. A region A represents the event: "the dart hit the region A." The probability of this event is the area of the region A. The union of the two circles in Fig. 1.2a represents the probability that either the events A, or the event B have occurred (this is the set of all points belonging to A or B, or both). The intersection of A and B in Fig. 1.2b represents the probability that both A and B have occurred (it contains all the points belonging to both A and B). Two disjoint events mean that there are no points belonging to both A and B, Fig. 1.2c. In this case, the probability of the event $A \cup B$ is simply the sum of the probabilities $P(A)$ and $P(B)$. In Fig. 1.2d the event B is contained in A, i.e. all points of B belong to A. The occurrence of B implies the occurrence of A. The occurrence of A does not necessarily imply the occurrence of B. In this case, the probability of B is smaller than the probability of A.

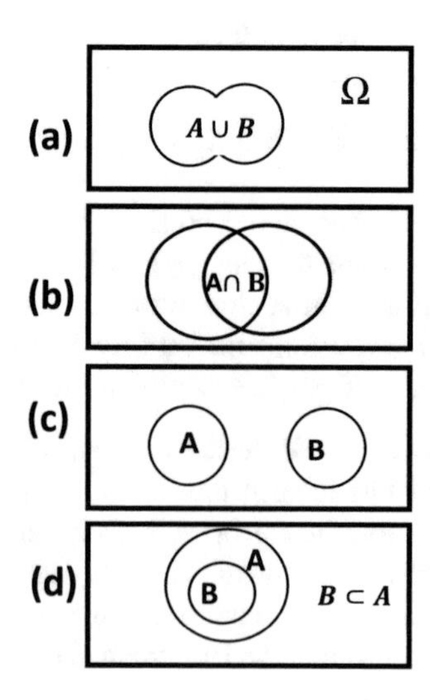

Fig. 1.2 Venn diagrams for **a** union of two events, **b** intersection of two events, **c** disjoint of two events, **d** inclusion of one event into another

It should be emphasized that the concept of disjoint event is defined in terms of the events, i.e. two events are said to be disjoint if, and only if the intersection between the two events is zero (or the empty set).

The concept of dependence (or independence) between two events A and B is defined not in terms of the events themselves but in terms of their probabilities. Two events are said to be independent if, and only if the probability of the intersection is equal to the product of the probabilities of the two events.

$$P(A \cap B) = P(A)P(B) \tag{1.10}$$

For example, if two people who are far apart from each other throw a fair die each, the outcomes of the two dice are *independent* in the sense that the occurrence, of say, "5" on one die does not have any effect on the probability of occurrence of a result, say, "3" on the other. On the other hand, if the two dice are connected by an inflexible wire, the outcomes of the two results could be dependent.

We define the extent of disjoint between the two events by the overlapping area in the Venn diagram. The extent of dependence is measured in terms of the correlation between the two events (assuming that *P*(*A*) and *P*(*B*) are non-zero):

$$g(A, B) = \frac{P(A \cap B)}{P(A)P(B)} \tag{1.11}$$

We say that the two events are *positively correlated* when $g(A, B) > 1$, i.e. the occurrence of one event enhances or increases the probability of the second event. We say that the two events are *negatively correlated* when $g(A, B) < 1$, and that they are *uncorrelated* or *indifferent* when $g(A, B) = 1$.

As we shall see below, there is no obvious relationship between the *extent* of dependence between two events, and the size of the overlapping events in the Venn diagram.

It has become common to represent the Mutual information (MI) between two random variables (random variables), which is a measure of the extent of dependence between two random variables by a Venn diagram, see Fig. 1.3.

The area of each circle is supposed to represent the *size* of the SMI while the intersection area is supposed to represent the extent of dependence between the two random variables. This, unfortunately is an unwarranted practice. It is unfortunate

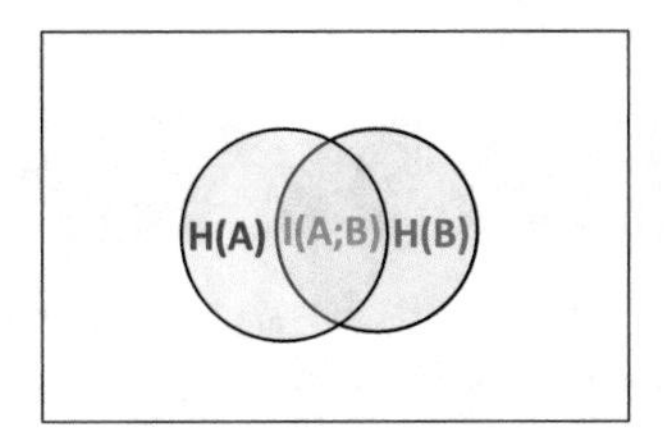

Fig. 1.3 A Venn diagram used for SMI: H(A) and H(B) and the MI, I(A; B)

that even some mathematicians who are supposed to know the meaning of a set of points (or events), and their measure (or the probability of an event), use such diagrams. An exception is Mackay [5], who warns that using the Venn diagrams might be "misleading". In my view, it is not only misleading, but totally unwarranted practice. First, because it is not clear what the points in the plan represent when we draw a Venn diagram for SMI. (Remember, the SMI is a single number not a set of points). Second, the intersection area which is supposed to represent the extent of dependence between two random variables turn out to be totally inappropriate. As will be shown below, the size of the overlapping areas is not proportional to the extent of dependence.

Finally, the most absurd outcome of using Venn diagram, occurs when one defines the MI between three random variables. We shall see that in some cases, the triple MI may have a *negative* value. In this case, one certainly cannot represent the extent of dependence by a *positive* intersection area in Venn's diagram.

In the following we show two examples of two dependent random variables. One uses two coins interacting through magnets. In this case, there is absolutely no analogue of the set of points in a Venn diagram. The second was designed intentionally to be a deceiving example. It makes use of a legitimate Venn diagram between two events which are represented by areas in a plan. We use this example to show that the *extent* of *overlapping* areas is not a monotonic function of the *extent* of *dependence.*

1.4.1 The Case of Two Coins with Magnets

Suppose that we have two coins, in each of which a little magnet embedded at its center (Fig. 1.4). We assume that the interaction between the two magnets are such that there is higher probability for the "up-down" pair, and lower probability for the "up-up" pair. The strength of the interaction between the magnets fall when the distance is very large.

Without getting into details of how the probabilities are calculated, it is clear that when we increase the distance R between the two tossed coins, the extent of dependence between the two coins will become increasingly smaller until we reach such a distance that the outcomes on the two tossed coins become independent. We choose the probabilities according to the rule:

Fig. 1.4 Two coins, in each of which a little magnet is embedded at its center. The magnet is perpendicular to the coin and points to the Head (H)

$$P(x_1, x_2) = \exp[-(x_1 x_2)/R] \Big/ \sum \exp\left[-\frac{x_1 x_2}{R}\right] \tag{1.12}$$

In this equation x_i represents the "state" of the magnet i; "either "up" or "down." The sum in (1.12) is over all $x_1, x_2 = 1, -1$, where 1 and -1 are assigned to the state "up" and "down" of the spin, or head (H) and tail (T) of the coin, respectively. For instance, at a unit distance ($R = 1$) the pair-probabilities are:

$$\begin{aligned} P(H, H) &= P(T, T) = 0.06 \\ P(H, T) &= P(T, H) = 0.44 \end{aligned} \tag{1.13}$$

When $R \to 0$, we have:

$$\begin{aligned} P(H, H) &= P(T, T) = 0 \\ P(H, T) &= P(T, H) = 0.5 \end{aligned} \tag{1.14}$$

Figure 1.5 shows how these probabilities change as a function of R. Clearly, when $R \to \infty$ all the four pair-probabilities tend to 0.25, i.e. each pair of outcomes has the same probability.

Figure 1.6 shows the SMI and the Mutual information (MI) for this experiment. As expected at $R = 0$, the SMI equal to 1 (there are two equally probable outcomes) and when R is very large the SMI tends to 2 (four equally probable outcomes). The MI starts at one for small R (maximum dependence), then drops to zero when R increases (zero dependence).

Obviously, there is no analog of the set of points in the plan which represents the elementary events. When we bring the two coins closer, we get more and more dependence, but there is no way to represent the extent of dependence by an area of overlapping regions in a plan.

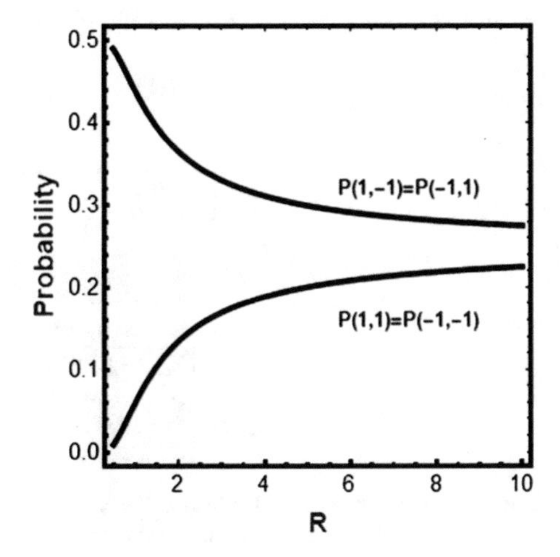

Fig. 1.5 The various pair-probabilities as a function distance R, for the two coins

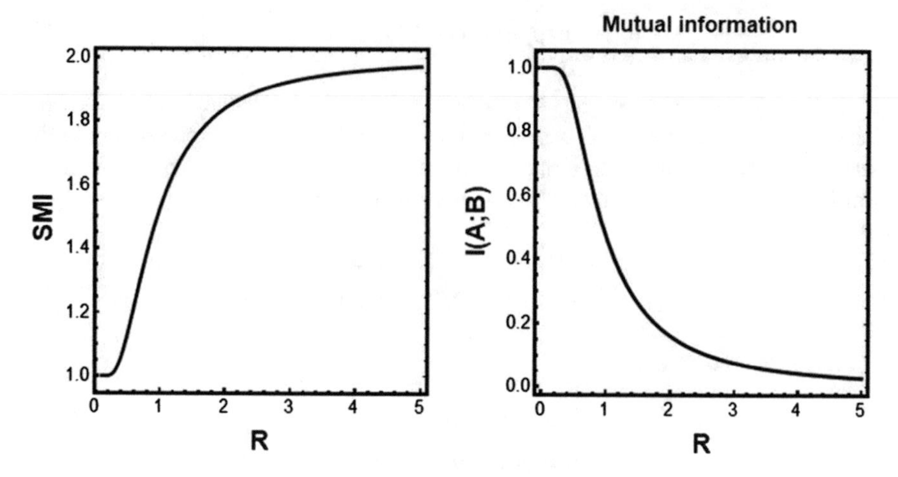

Fig. 1.6 The SMI and the MI for the two coins as a function of the distance R

1.4.2 The Case of Two Regions on a Board at Which a Dart Is Thrown

We discuss here an example, for which Venn diagram may be used for representing the various probabilities, but cannot be used to measure the extent of correlation (between two random variables).

We start with a simple case discussed in details in Ben-Naim [3]. We have a board of unit area. On this board, we draw two regions A and B, of equal areas, Fig. 1.7.

We throw a dart on this board. Clearly, since we chose the area of the entire board as unity, the probability of finding the dart (thrown blindly at the board) at any given region is equal to the area of that region. If the area of each rectangle in Fig. 1.7 is $q = 0.1$, then the probability of finding the dart at one specific rectangle is:

$$P(A) = P(B) = q = 0.1 \tag{1.15}$$

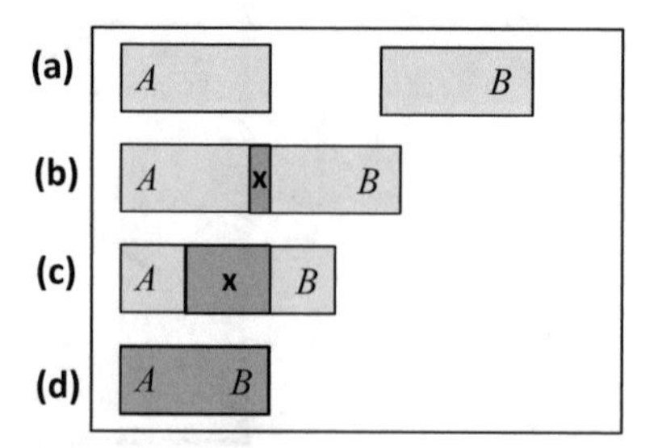

Fig. 1.7 Two regions A and B having the same area q. **a** Extreme negative correlation, **b** negative correlation, **c** positive correlation, and **d** extreme positive correlation

Next, we bring the two rectangles closer and closer to each other. We denote by x the overlapping area between A and B. The joint probability of finding the dart in *both* A and B is:

$$P(A \cdot B) = x \tag{1.16}$$

The correlation between the two events A and B is defined by:

$$g(A, B) = \frac{P(A \cdot B)}{P(A)P(B)} = \frac{x}{q^2} \tag{1.17}$$

When $x = 0$, the two events are *disjoint* (A and B have no common point), Fig. 1.7a. This means that the two events are negatively correlated; knowing that A occurred, *excludes* the occurrence of B. We call this negative correlation since in this case:

$$\begin{aligned} P(A|B) - P(A) &= \frac{P(A \cdot B)}{P(B)} - P(A) \\ &= P(A)g(A, B) - P(A) = P(A)(g(A, B) - 1) < 0 \end{aligned} \tag{1.18}$$

Thus, negative correlation means that the conditional probability $P(A|B)$ is smaller than the probability $P(A)$. When x is very small, say $x = 0.001$, see Fig. 1.7b, the correlation is still negative: $g(A, B) = 0.001/0.01 = 0.1$.

When x increases from zero to q, the quantity $P(A|B) - P(A)$ changes from negative to positive. There is an intermediary point, $x = q^2$, when $g(A, B) = 1$. This is the case of "no correlation," for which we have:

$$g(A, B) = \frac{q^2}{q^2} = 1 \tag{1.19}$$

When $x > q^2$ the correlation is positive, see Fig. 1.7c.

At $x = q$, we have maximum overlapping of A and B, and also maximum positive correlation Fig. 1.7d. The value of $g(A, B)$ is:

$$g(A, B) = \frac{q}{q^2} = \frac{1}{q} > 1 \tag{1.20}$$

In measuring the extent of dependence by the MI, we use $\log_2 g$ as a measure of dependence. In this case, negative correlation is when $g < 1$(or $\log g < 0$), positive correlation when $g > 1$(or $\log g > 0$), and no correlation when $g = 1(\log g = 0)$. Plotting $\log g$ as a function of x, we find that $\log g$ changes from *negative* to *positive* values at $x = 0.01$ (here, we chose $q = 0.1$). This finding already indicates that one cannot use the extent of overlapping (x) to represent the extent of correlation ($\log g$), or the MI, see below.

In this example, we use two *events* A and B, and the correlation between the two events, $g(A, B)$. to measure the extent of dependence.

Next, we use the same game or experiment as in Fig. 1.7, but we use two *random* variables A and B, and we use the *mutual information* as a measure of the extent of dependence between the two random variables.

The experiment is as in Fig. 1.7, but here A and B are random variables or two games. The random variable A gets two values: zero and one, for being empty or occupied by the dart, respectively. The probability distributions of A and B are:

$$P(A = 1) = q, \ P(A = 0) = 1 - q \tag{1.21}$$

$$P(B = 1) = q, \ P(B = 0) = 1 - q \tag{1.22}$$

The corresponding SMI of A and B are:

$$SMI(A) = SMI(B) = -q \log q - (1 - q) \log(1 - q) \tag{1.23}$$

The pair-probability distribution for A *and* B is:

$$\begin{aligned} &P(A = 1, B = 1) = x \\ &P(A = 1, B = 0) = P(A = 0, B = 1) = (q - x) \\ &P(A = 0, B = 0) = 1 - 2q + x \end{aligned} \tag{1.24}$$

The corresponding $SMI(A, B)$ is:

$$\begin{aligned} SMI(A, B) &= -\sum_{i,j=0,1} P(i, j) \log P(i, j) \\ &= -x\text{log}x - 2(q - x) \log(q - x) - (1 - 2q + x)\text{log}(1 - 2q + x) \end{aligned} \tag{1.25}$$

The MI between A and B is calculated from the definition:

$$\begin{aligned} I(A; B) &= \sum_{i,j=0,1} p(i, j) \log g(i, j) \\ &= SMI(A) + SMI(B) - SMI(A, B) \end{aligned} \tag{1.26}$$

Figure 1.8 shows the pair SMI. For the choice of $q = 0.1$ (i.e. the area of each rectangle is 0.1) as a function of x, between $x = 0$ and $x = q = 0.1$ (i.e. from zero overlapping to total overlapping). The $SMI(A, B)$ starts at about 0.92 when $x = 0$, and ends at about 0.47 when $x = q$, (not shown in this figure). Between these two values there is a *maximum* at $x = q^2 = 0.01$.

Figure 1.9b shows MI for the pair A, B. The most interesting aspect of this curve is that it passes through at minimum at $x = q^2 = 0.01$. $I(A; B)$ must be positive, by

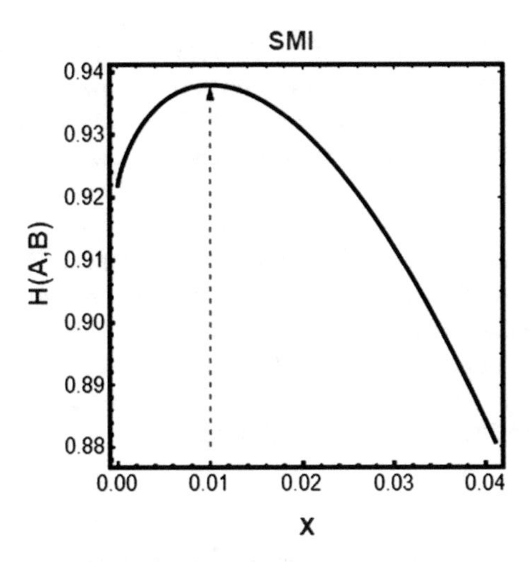

Fig. 1.8 The pair-SMI, H(A, B) as a function of the overlapping regions A and B

definition (see Chap. 3 of Ben-Naim [1]), however the MI is an average over all four pair correlations, each of which can be either positive or negative, see Fig. 1.9a. In Fig. 1.9c, we show the derivative of $I(A; B)$ with respect to x.

The value of the MI starts as about $I(A; B) \approx 0.016$ at $x = 0$. It then decreases to zero at $x = 0.01$, then increases to about 0.47 when $x = q$ (total overlapping of the regions A and B).

Clearly, when x changes from zero to q, the overlapping regions increase monotonically (from zero to 0.1), but the value of $I(A; B)$, measuring the extent of dependence behaves differently; first, it goes down, passes through a minimum, then goes up. Clearly, one *cannot* represent the extent of dependence by the overlapping areas between two regions in a plan.

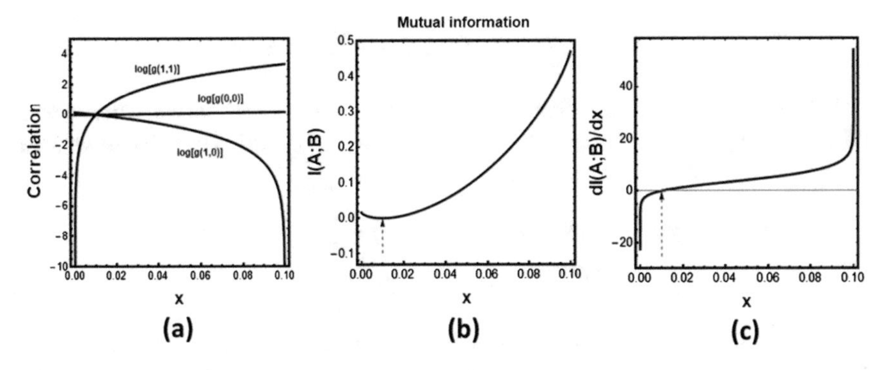

Fig. 1.9 **a** All four (log) pair correlations between A and B. **b** The MI between A and B, and **c** the derivative of I(A; B) with respect to x

Conclusion: One should not confuse the extent of overlapping of events with the extent of dependence between events. We have shown that for a *pair* of *events* one can use the Venn diagram. However, one cannot use such diagram for a pair of *random variables*. We provide some more details in the following sections and in the Appendices.

In the next section, we shall show that this conclusion is also valid for three events. We shall see that three events might have a common intersection area, yet the three events are triple-wise independent. Also, the three events might be triple-wise disjoint (i.e. no common points belonging to the three events), yet there is triple-wise dependence. Furthermore, some of the triple-MI might be negative. Hence, the extent of dependence between three events (as well as three random variables) cannot be represented by a (positive) area in Venn diagram.

1.5 The Frustrating Search for a Measure of Frustration

In Chap. 4 of Ben-Naim [1], we introduced the concept of frustration in systems of three or more spins. We also discussed the various possibilities of generalizations of the concept of mutual information (MI). We saw that the most *natural* generalization of the MI is the quantity which we referred to as the *total correlation*, which for n random variables is defined by:

$$TI(X_1; \dots; X_n) = \sum_{x_1, \dots, x_n} P(x_1, \dots, x_n) \log g(x_1, \dots, x_n) \tag{1.27}$$

where $g(x_1, \dots, x_n)$ is the correlation between the n events $X_1 = x_1, X_2 = x_2, \dots X_n = x_n$.

There are other definitions of MI which we referred to as *conditional* MI. For three random variables, this is defined by:

$$CI(X_1; X_2; X_3) = I(X_1; X_2) - I(X_1; X_2|X_3) \tag{1.28}$$

This is a measure of the difference between the pair MI, and the *conditional* pair MI, given the presence of a third rv. It turns out that this definition can also be written in terms of the SMIs in this system:

$$\begin{aligned} CI(X_1; X_2; X_3) = H(X_1) + H(X_2) + H(X_3) - H(X_1; X_2) \\ - H(X_1, X_3) - H(X_2, X_3) + H(X_1, X_2, X_3) \end{aligned} \tag{1.29}$$

The sum of the terms on the right-hand-side of Eq. (1.29) is reminiscent of the inclusion–exclusion principle in probability (recall that $H(X_1)$ is the SMI of X_1). This fact has led some authors to use Venn diagrams to represent the extent of dependence

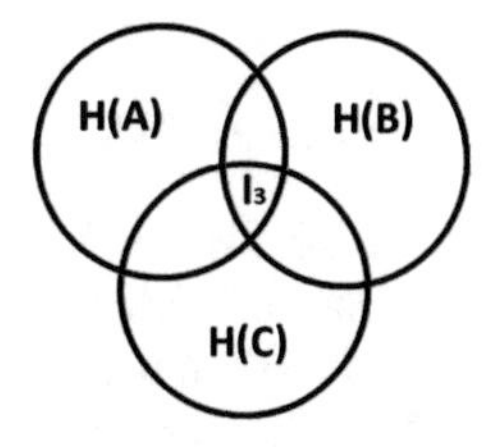

Fig. 1.10 Venn diagram used for the three SMI and triplet mutual information, denoted by Matsuda as I_3, and in the present book as CI(A; B; C)

between three random variables. A typical such a Venn diagram is shown in Fig. 1.10 [the triplet MI is denoted I_3 by Matsuda [14]].

Unfortunately, this is a misleading representation of the MI. As we have seen in the previous section, Venn diagrams cannot be used to represent the extent of dependence between two random variables. It is a fortiori misleading to do so for three or four random variables. In this section, we shall examine a few examples for which there is no relationship between the extent of overlapping between three events (or three random variables) and the extent of dependence. In fact, we shall see examples where the conditional MI between three events could be *negative*. This certainly cannot be represented by a (positive) overlapping area in a Venn diagram.

The conditional MI has also been interpreted as a measure of frustration [14]. As we have seen in Ben-Naim [1], there are cases where the CI turns out to be negative in a three-spin system. However, this is not always true. We shall present a few examples where CI can be either positive or negative, but it is impossible to relate these values to extent of frustration in the system. We will conclude that the quantity CI cannot, in general, be used as a measure of frustration.

1.5.1 Three Coins with Magnets

The first example is an extension of the example discussed in Sect. 1.4.1. Instead of two coins, we have three coins each having a magnet, or a spin, at its center so there are interactions between the magnets. The centers of the three coins form a perfect triangle with edge R. We assume that the interaction energy between the three magnets has the form:

$$U(x_1, x_2, x_3) = (x_1x_2 + x_1x_3 + x_2x_3)\big/ R \tag{1.30}$$

where x_1 and x_2 can have the values of 1 and -1 corresponding to the states of the magnet: "up" and "down," respectively. Clearly, whenever both x_1 and x_2 have the same sign, we have a positive interaction energy, and when they have different signs we have negative interaction energy. All the probabilities in this system are derived from the equation:

$$P(x_1, x_2, x_3) = \frac{\exp[-U(x_1, x_2, x_3)]}{\sum_{x_1,x_2,x_3} \exp[-U(x_1, x_2, x_3)]} \tag{1.31}$$

Note that we use here the Boltzmann distribution, with $(k_B T) = 1$. This is very similar to the three spin system we have discussed in Chap. 4 of Ben-Naim [1], and also in Chap. 3 of this book. The only difference is that here we are not interested in the temperature dependence of the various quantities, but only to the extent of interaction, hence extent of dependence between the spins—which varies with the distance R.

There are altogether eight possible configurations of the three spins as shown in Fig. 1.11. For any distance R, we have two high-probability configurations (either all "up-up" or all "down-down") and six configurations with lower probability (one "up" and two "down," or one "down" and two "up").

In Fig. 1.12b, we show the pair-probabilities for this system as a function of the distance R. One should compare this figure with Fig. 1.5, which is reproduced in Fig. 1.12a, for two coins. Note that $P(1, 1)$ in Fig. 1.12b is the probability of finding the pair of two coins in a state "up-up" in the presence of the third coin (we sometimes denote this probability by $P(1, 1, \_)$ which means "up-up-unspecified").

As for the case of two coins when $R \to \infty$, there is no dependence, hence, all four configurations of the two coins will have the same probability of 0.25. This is true also in the case of three coins; when R is very large, the probability of any configuration of the pair is independent of the third coin.

The behavior is much different in the case of very small R. As we saw in Fig. 1.5, which we reproduce here in Fig. 1.12a, at $R \to 0$ we had two configurations with probability 0.5 ("up-down" and "down-up), and two configurations with zero probability ("up-up" and "down-down").

The behavior is different when we have three coins, Fig. 1.12b. When $R \to 0$, we have two configurations with probability of 0.333 ("up-down" and "down-up"), and two configurations with a lower probability of 0.166 but not zero as in Fig. 1.12a ("up-up" and "down-down").

The reason for this different behavior of the pair probabilities is the presence of a third coin. The reader should pause to think and understand these results; this understanding will be essential in understanding the effect of a third coin on both the pair correlation and the pair MI.

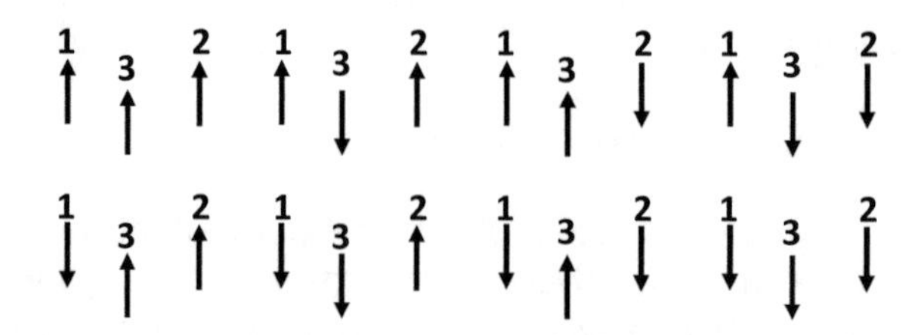

Fig. 1.11 All eight configurations of the three spins; 1, 2, and 3

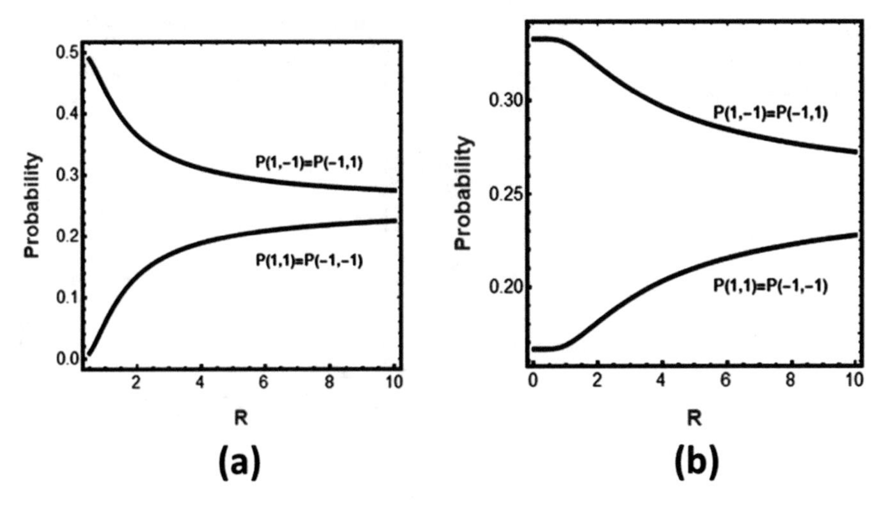

Fig. 1.12 The various pair-probabilities as a function distance for the two coins. **a** For pair of coins, reproduced from Fig. 1.5, **b** for pair of coins but in the presence of a third coin

Remember that when R is very small, the interactions are very strong. When we have only two coins, the strong attractive interaction between the two magnets *determines* the eventual probabilities of 0.5, 0.5 (for "up-down" and "down-up"). In the presence of a third coin, the situation is very different. Here, at very short distance, the probability of the configurations "up-down" and "down-up," is not determined only by the interaction between the two coins, but also on the interactions of the two coins with the third coin, these interactions are also very *strong* at this distance.

Similarly, for two coins at $R \to 0$, the configuration "up-up" has zero probability. However, when a third coin is present, although its configuration is not specified, it has an influence on the probability of the configuration of the pair. For instance, the "up-up" configuration of one pair will have non-zero probability when the third coin is "down."

This can be better understood by examining the triplet-probabilities, Fig. 1.13. At $R \to \infty$, all eight configurations become equally probable (1/8). When $R \to 0$, there are two configurations with zero probability "all-up" and "all-down," and all the other six configurations have equal probability: $\frac{1}{6} = 0.166$. When we calculate the pair-probability, say of $P(1, 1, \_)$ we should remember that it is the marginal probability:

$$P(1, 1, \_) = P(1, 1, 0) + P(1, 1, 1) = 0.1666 + 0.0 \tag{1.32}$$

Here, we clearly see the origin of the non-zero probability of the pair "up-up." It is because the probability of the triplet configuration $(1, 1, 0)$ is 0.166. Whereas in the two-coin case this effect of the third coin is absent.

The effect of a third coin on the pair-probability also affects the pair correlations. For instance, the pair correlation, $g(1,1)$ is defined by:

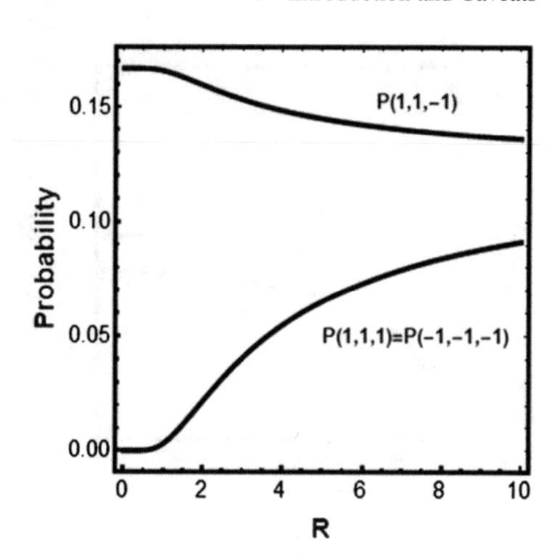

Fig. 1.13 The triplet-probabilities for the three coins as a function distance

$$g(1,1) = \frac{P(1,1)}{(P(1))^2} \tag{1.33}$$

For the two-coin system we have:

$$g(1,1) \to 1, \text{ for } R \to \infty \text{ (no correlation)}$$
$$g(1,1) \to 0, \text{ for } R \to 0 \text{ (infinite negative correlation)}$$

whereas in the three-coin system we have:

$$g(1,1) \to 1, \text{ for } R \to \infty \text{ (no correlation)}$$
$$g(1,1) \to 0.66, \text{ for } R \to \infty \text{ (finite negative correlation)}$$

Figure 1.14 shows the pair and the triplet SMI for the three-coin system. As expected both start at a lower value at $R \to 0$. The actual values of SMI at $R \to 0$ is determined by the probability distributions of the pairs and the triplets. The values of the pair and the triplet SMI at $R \to \infty$ are 2 and 3, as expected (due to four and eight equally probable configurations, respectively).

Next, we show in Fig. 1.15 the pair MI as a function of R. The general behavior is similar to the case of the two-coin system. Compare with Fig. 1.6. In both cases, $I(A;B) \to 0$ at $R \to \infty$ (independence). However, there is a great difference in the limit of $R \to 0$. For the two-coin system we saw that $I(A;B) = 1$ at $R \to 0$, due to fact that:

$$I(A;B) = 2H(A) - H(A,B) \to 1 + 1 - 1 = 1 \tag{1.34}$$

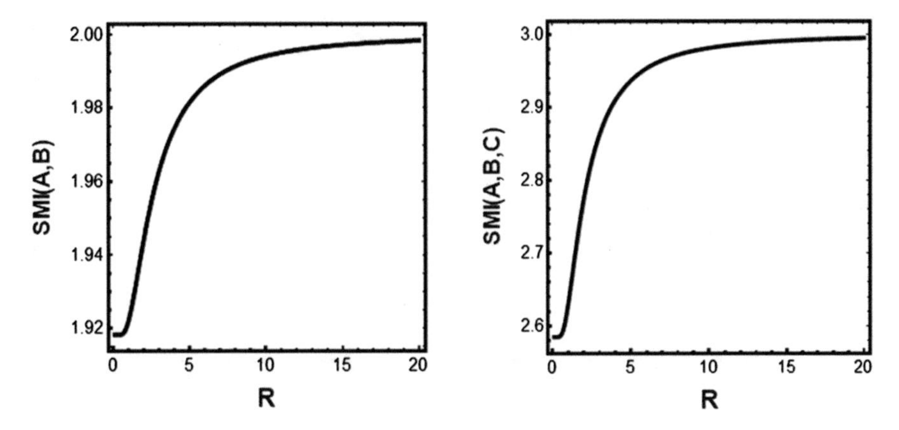

Fig. 1.14 The pair and the triplet SMI as a function of the distance R

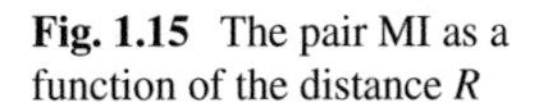

Fig. 1.15 The pair MI as a function of the distance R

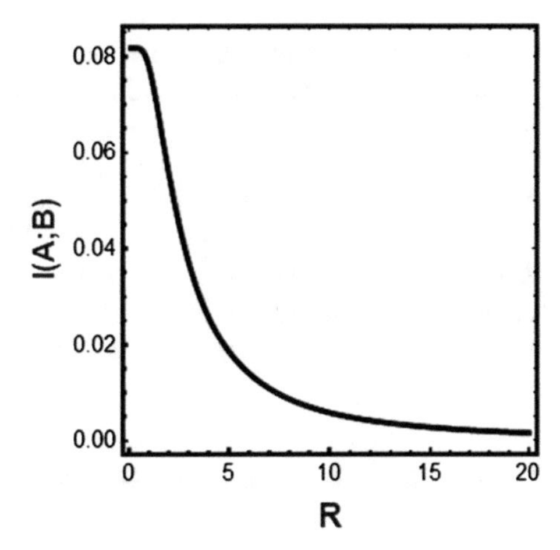

In the case of the three-coin system $I(A; B)$ is only 0.08 at $R \to 0$. In this case, the value of the MI is obtained from:

$$I(A; B) = 2H(A) - H(A, B) \to 1 + 1 - 1.92 = 0.08 \tag{1.35}$$

Clearly, this large effect of the third coin on the pair MI is due to the strong interaction between the third coin and the pair of coins A and B.

Figure 1.16 shows this total and the conditional triplet MI. The total triplet MI is always positive as expected. However, the conditional triplet MI is everywhere negative. It is relatively large negative at $R \to 0$, and tends to zero when $R \to \infty$. As we have discussed in Chap. 4 of Ben-Naim [1], the negative values of CI were interpreted as arising from frustration in the system. As we shall see in Chaps. 3 and 5, it is not always true that negative CI is associated with frustration. Here,

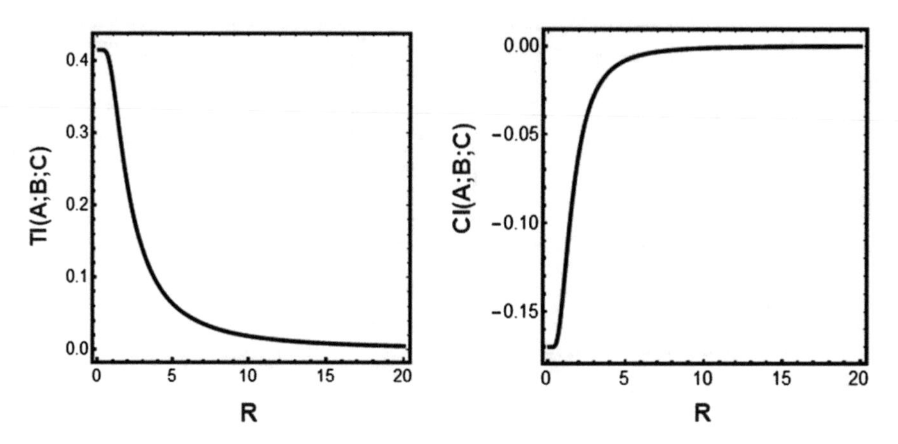

Fig. 1.16 The total and the conditional MI as a function of the distance R

we want to emphasize that the negative CI is the result of the definition of CI as a difference between $I(A;B)$ and $I(A;B|C)$. It is by no means clear that when $I(A;B|C)$ becomes larger than $I(A;B)$, there will be frustration in the system. In the next examples we shall see that the CI may be negative even though there is no frustration in the system. One final comment on this result follows from the definition of CI in terms of the SMI (see Eq. 1.29) which led (or rather misled) people to use Venn diagram for the CI. It is clear that in this example one cannot represent SMIs and MI as areas in a Venn diagram. The fact that CI might be negative makes its representation by a positive area in a Venn diagram inappropriate.

1.5.2 Three Regions on a Board

In the example of Sect. 1.5.1 we had three coins, or two spins, each of which could be in one of two states, "up" or "down." We saw that there is no way of representing either the SMI or the MI in a Venn diagram.

In the next example we replace the three coins by three regions on a board. We throw a dart on the board of unit area. We know that the dart hit the board. The events are: "the dart is in region A" (or B, or C). We shall treat this system in two languages. First, as events having probabilities and represented in a Venn diagram. Second, as random variables, having SMIs and MIs which cannot be represented by a Venn diagram.

The system discussed in this section is shown in Fig. 1.17.

It is an extension of the system discussed in Sect. 1.4.2. Instead of two overlapping regions, we have here three overlapping regions, only in pairs, not in triplets. We assume that a dart was thrown on a board of unit area. Each of the regions A, B, and C have the same area chosen as $q = 0.1$, hence, the probability of finding the dart in any of these areas is 0.1.

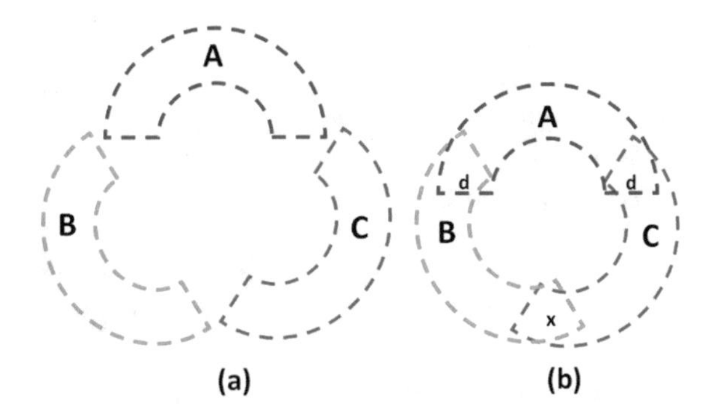

Fig. 1.17 **a** Three non-overlapping regions, **b** the same with overlapping in pairs only

We denote by d the area of overlapping between A and B, and between A and C. We denote by x the overlapping area between B and C. We start by listing the triplet probabilities which can be read from Fig. 1.17, These are:

$$P(1,1,1) = 0 \text{ (no triple overlapping)}$$
$$P(0,0,0) = 1 - 3q + 2d + x$$

(this is the area of the whole board minus the area of $A \cup B \cup C$)

$$\begin{aligned} P(1,0,0) &= q - 2d \\ P(0,1,0) &= q - d - x \\ P(0,0,1) &= q - d - x \\ P(1,1,0) &= d \\ P(1,0,1) &= d \\ P(0,1,1) &= x \end{aligned} \tag{1.36}$$

Clearly, the sum of all these is one:

$$\sum_{x_1, x_2, x_3} P(x_1, x_2, x_3) = 1 \tag{1.37}$$

Here, each of x_i may either be 1 or 0 (occupied or empty). All the pair and singlet probabilities are derived from (1.49). Specifically, the pair-probability $p(1, 1, \_)$ is defined by:

$$P_{AB}(1,1) = P_{AC}(1,1) = P(1,1,0) + P(1,1,1) \tag{1.38}$$

This is different from $P_{BC}(1, 1)$, which is defined by:

$$P_{BC}(1,1) = P(\_,1,1) = P(0,1,1) + P(1,1,1) \tag{1.39}$$

Figure 1.18 shows these pair probabilities as a function of d, and a fixed value of $x = 0.01$. As expected, $P_{AB}(1, 1)$ and $P_{AC}(1, 1)$ increase linearly with d, but independently of x, whereas $P_{BC}(1, 1)$ is independent of d, and is equal to $x(x = 0.01$, in this figure).

The singlet probabilities are:

$$\begin{aligned} P_A(1) &= P_B(1) = P_C(1) = 0.1 \\ P_A(0) &= P_B(0) = P_C(0) = 0.9 \end{aligned} \tag{1.40}$$

Next, we discuss the SMI for this section. The singlet SMIs are equal and have a constant value independent of d, and x. These are:

$$\begin{aligned} SMI(A) &= SMI(B) = SMI(C) \\ &= -P_A(1)\log P_A(1) - P_A(0)\log P_A(0) = 0.47 \end{aligned}$$

The pair SMI, for the pair BC is independent of d, but SMI(A, B) = SMI(A, C) depends on d (but is independent of x). Figure 1.19a shows that this SMI goes through a maximum. The triplet SMI(A, B, C) has a similar form and is shown in Fig. 1.19b.

Next, we turn to the pair MI. Figure 1.20 shows the pair-MI for AB (equal to AC) and the pair BC. The former passes through a minimum at $d = 0.01$, whereas the latter is independent on d, but has a constant value for different x.

The values of $I(A; B)$ in Fig. 1.20 should be compared with the values of $I(A; B)$ for two overlapping areas shown in Fig. 1.19b, Sect. 1.4.2. Note that both have the same form, but the values of the MI are much smaller for the pair of AB in the *presence* of C, than for the case of the absence of C.

Figure 1.21 shows the *total* triplet MI. As expected, the values are all positive. Note that for each value of d, the MI first decreases with x, then increases when x increases. For a fixed x, the MI goes through a minimum value.

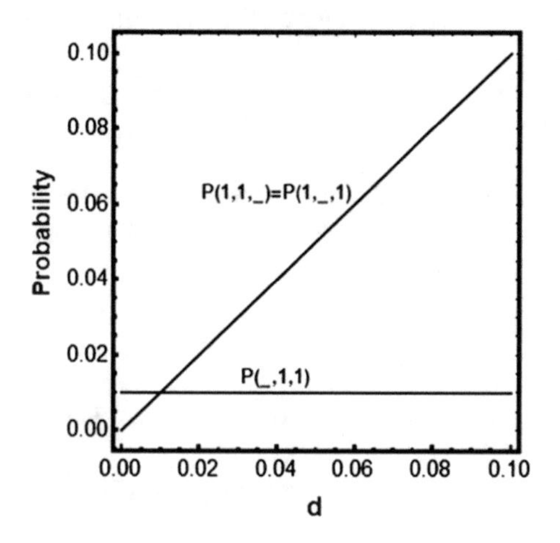

Fig. 1.18 The various pair-probabilities as a function distance d, and a fixed value of x

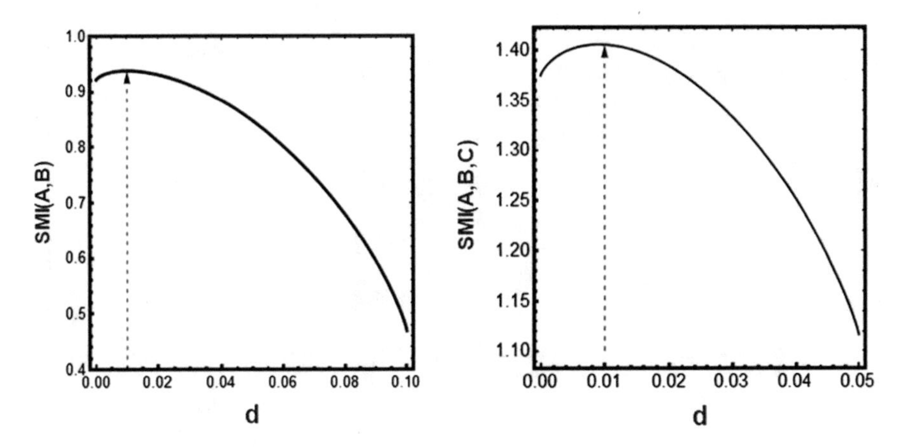

Fig. 1.19 The pair and the triplet SMI as a function of the distance *d*

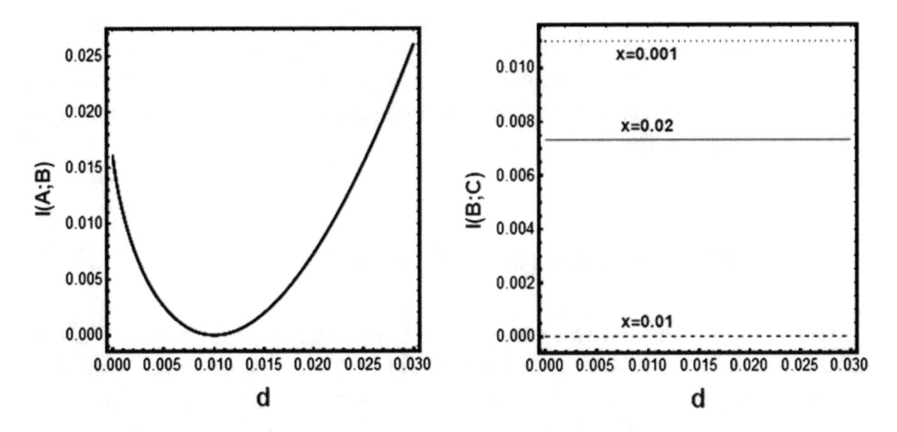

Fig. 1.20 The pair-MI, I(A; B) and I(B; C) as a function of the distance *d*

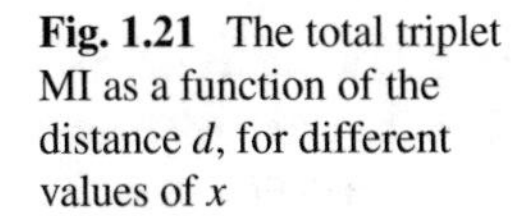

Fig. 1.21 The total triplet MI as a function of the distance *d*, for different values of *x*

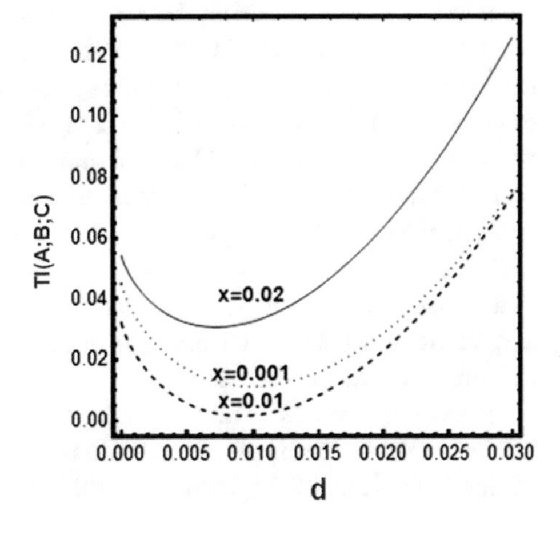

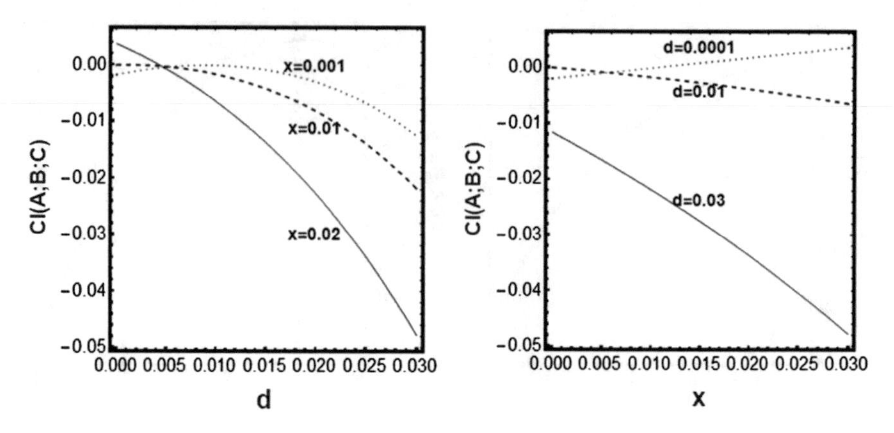

Fig. 1.22 The conditional triplet MI as a function of the distance d, and as a function of x

The most interesting results are the values of the conditional triplet MI, Fig. 1.22. Here, we get both positive and negative values of the CI when we change the distance d. It is clear from these results, that there is no reason to interpret the negative value of CI as a measure of frustration.

1.5.3 A Caveat to the Caveat on Frustration

In this section, we showed two examples of three-random variables for which we found negative values of the triple-conditional MI. In the first example (coins with embedded magnets) we have frustration as defined for a three-spin system, and discussed in Chap. 4 of Ben-Naim [1], However, there is not even a hint as to how the values of SMI may be represented as areas in a Venn diagram, and the MI as overlapping areas measuring the extent of dependence.

In the second example (with areas on a board on which a dart hit) we can describe the various *events* on a Venn diagram, but this description is valid only when we treat *events* and their probabilities, not when we treat random variables and the dependence between them. Once we move from events to random variables, and the corresponding SMI and MI, there is no way to describe the SMI and the extent of dependence on a Venn diagram.

The second question is whether or not a negative value of CI may be considered as a measure of frustration in the system (without any reference to a description by a Venn diagram). The answer to this question depends on which of the (equivalent) definition we use for the CI.

In general, we can safely say that none of the definitions of the CI offer any interpretation as a measure of frustration. Again, one should be careful here about the distinction between the treatment on the "level" of single events, their probabilities, and the extent of dependence on one hand, and the treatment on the "level" of random variable, their SMIs, and the extent of dependence, on the other hand.

Consider the following story which might be interpreted as frustration on one level, but not in general, on the other level. [A more detailed story may be found in Ben-Naim [9]].

Suppose a crime was committed in area A as in Fig. 1.17. A suspect S was apprehended and was charged with the crime. Police officer P_B said that he saw S in the area B at the same time when the crime was committed. If the overlapping area between A and B is large enough, then the fact that S was seen in B *supports* the claim that he was also in A ("supports" here in the sense that $g(A; B) > 1$). Similarly, suppose that a second police officer P_C saw S in a different area C, which also overlaps with A to the extent that given that S was seen in C *supports* the claim that he was also in A. Thus, each of the witnesses *separately* support the claim that S was in A at the time of the crime. However, if the areas of B and C overlap (i.e. $x \neq 0$ in Fig. 1.17b), then the fact that S was seen in both C and B (i.e. in the region x) excludes the possibility that S was in A at the time of the commission of the crime.

Note that this story is on the "level" of events and their probabilities. One can say that both police officers P_B and P_C were *frustrated*; initially, each police officer believed that his evidence was supportive, but they must have been very "frustrated" to learn that their combined evidence was un-supportive. This is true from the point of view of the police enforcers. What about the suspect? Initially, the witnesses of each police officer separately seemed to have supported the claim that the suspect was present in the area A at the time of the commission of the crime. However, when both witnesses of police officers P_B and P_C arrived, the suspect must have been happy, and not frustrated at all. Now let us look at the same story from the point of view of information theory. The most relevant definition of CI for this story is:

$$CI(A; B; C) = I(A; B) + I(A; C) - I(A; (B, C)) \tag{1.41}$$

A negative value of CI means that knowing the joint rv (B, C) gives more information on A (the last term on the right hand side of 1.54), than the sum of the information given by B on A, and the information given by C on A. Can one interpret this negative sign of CI as frustration? I doubt it. To the best of my knowledge no other (equivalent) definition of the triplet MI has any hint to offer an interpretation of negative CI in terms of frustration in the system.

Finally, I would like to suggest to the reader to do the following exercises. Their detailed study should be beneficial in clarifying the two "levels" of treatment of each case, the relationship between overlapping events and their probabilities, on one hand, and independence between the random variables, and their MI on the other hand. Finally, examine the relationship (if any) between CI and frustration. The answers are given in the Appendices.

Exercise 1.1 Consider the same problem as in Sect. 1.5.2 but with a different arrangement of the three areas on the board, Fig. 1.23. Each region is a perfect square of edge $q = 0.1$, hence the area of each square is $q^2 = 0.01$. xq is the overlapping area

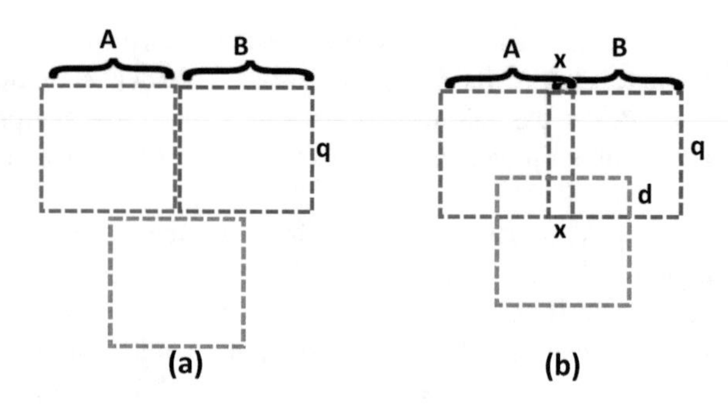

Fig. 1.23 The three regions for Exercise 1.1: **a** no overlap, **b** overlapping between all three regions

between A and B, and dq is the overlapping area between C and $A \cup B$. Calculate the triple probabilities in this system, the marginal probabilities, the various SMI and the pair and triplet MI. See solution below.

Solution to Exercise 1.1

The triplet probabilities may be read from the Fig. 1.23. The total area of the board is unity. The area of each square is q^2, and we chose $q = 0.1$, hence the probability of hitting any of the areas is $q^2 = 0.01$. The eight triplet probabilities are:

$$\begin{aligned}
P(1,1,1) &= xd \\
P(0,0,0) &= 1 - 3q^2 + qx + qd \\
P(1,0,0) &= q^2 - qx - d(q-x)/2 \\
P(0,1,0) &= q^2 - qx - d(q-x)/2 \\
P(0,0,1) &= q^2 - qd \\
P(0,1,1) &= d(q-x)/2 \\
P(1,0,1) &= d(q-x)/2 \\
P(1,1,0) &= x(q-d)
\end{aligned} \tag{1.42}$$

Check that the sum is:

$$\sum_{x_1,x_2,x_3} P(x_1,x_2,x_3) = 1 \tag{1.43}$$

All the other probabilities may be obtained as marginal probabilities from $P(x_1,x_2,x_3)$. We show in Fig. 1.24 the pair probability:

$$\begin{aligned}
P(A,C) = P(B,C) &= P(1,_,1) \\
&= P(_,1,1) = P(0,1,1) + P(1,1,1)
\end{aligned}$$

As expected $P(A,C)$ starts at zero when $d = 0$ and increases linearly with d.

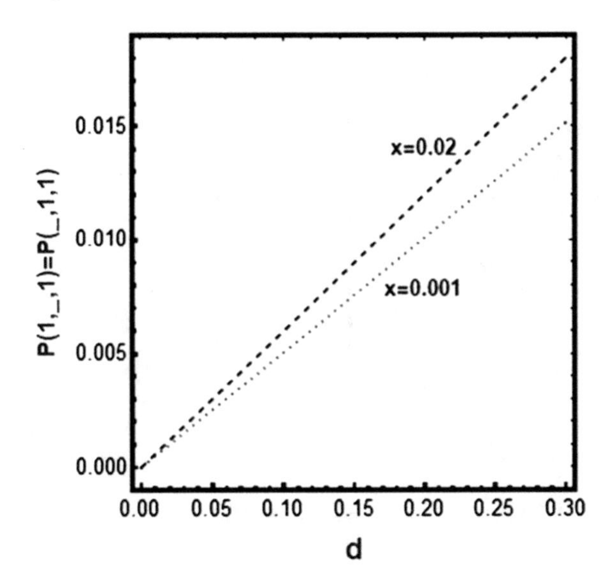

Fig. 1.24 The pair-probabilities as a function distance d

Figure 1.25 shows both SMI(A, C) and SMI(A, B, C). In both curves there is a maximum at a very small d, then the SMI decreases.

Figure 1.26 shows the pair MI, $I(A; B)$ and $I(A; C) = I(B; C)$ as a function of d, for different values of x. As expected $I(A; B)$ is independent of d, but is larger the larger overlapping between A and B. Note that both $I(A; C)$ and $I(A; B)$ goes through a minimum at which the value of the MI is zero. This is interesting because

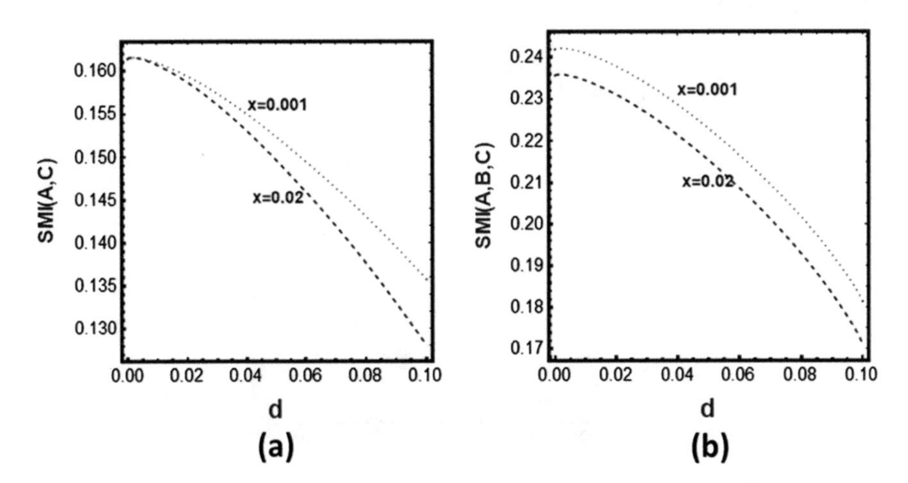

Fig. 1.25 Thea pair and the triplet SMI as a function of the distance d

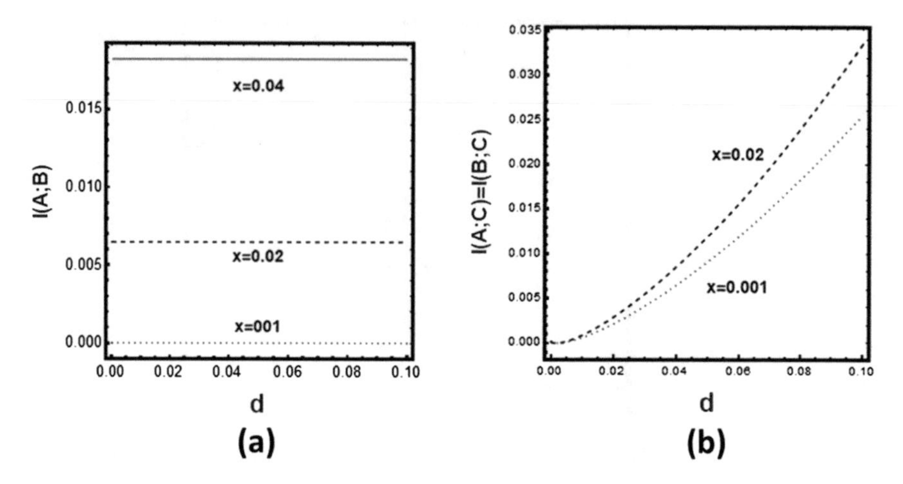

Fig. 1.26 The pair-MI, I(A;B) and I(A;C) = I(B;C) as a function of the distance d

initially when $d = 0$, there is dependence between A and C, and between A and B, then the extent of dependence drops to zero before raising again.

Figure 1.27 shows the triple total MI,$TI(A; B; C)$. As expected, the values are all positive. The curve for each x goes through a minimum as a function of d.

Figure 1.28 shows the conditional MI; as always this quantity is the most interesting. It has both positive and negative values. As we have noted in Sect. 1.5.2, it is not clear how to relate the negative values of CI with any frustration in the system. As in the example discussed in Sect. 1.5.2, also here we have overlapping regions in the Venn diagram for the three *events A*, *B*, *C* but this Venn diagram cannot be used for the random variables *A*, *B*, *C*.

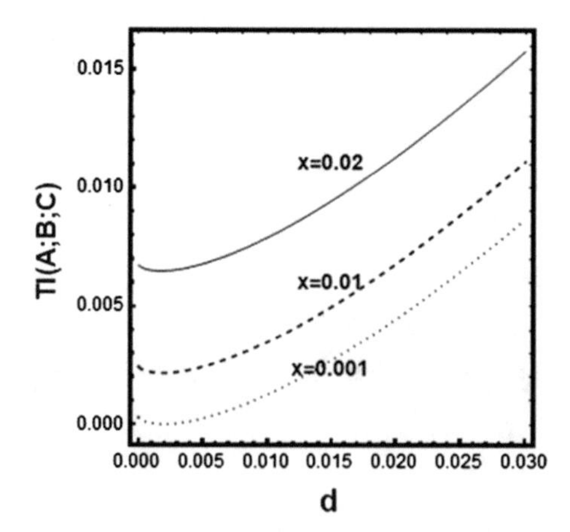

Fig. 1.27 The total triplet MI as a function of the distance d, for different values of x

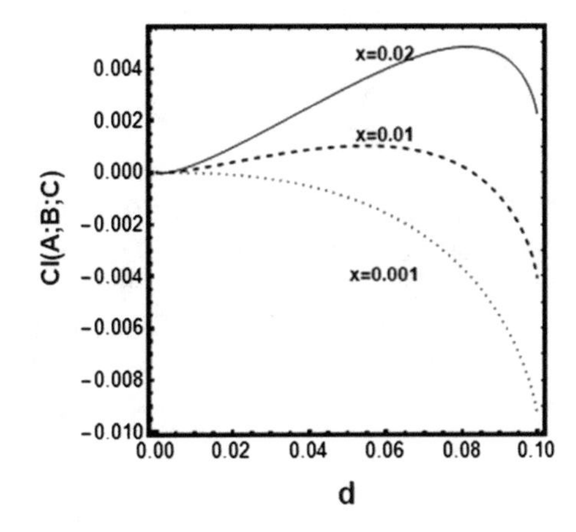

Fig. 1.28 The conditional triplet MI as a function of the distance d, for different values of x

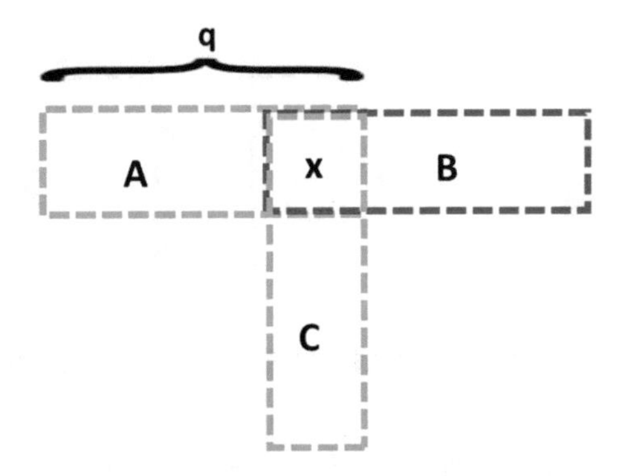

Fig. 1.29 The three regions for Exercise 1.2. The area of the overlapping region between the three rectangles is x

Exercise 1.2 This exercise was first discussed in Ben-Naim [9], in connection with the relationship between extent of overlapping of events and the extent of dependence. The arrangement of the three regions on a board is shown in Fig. 1.29.

The description of the same system is the same as the one discussed in Sect. 1.5.2. The triplet probabilities may be read from the figure. Calculate the SMI, the pair and the triplet MI for this system. Can you make any conclusion regarding the connection between the values of the conditional triplet MI, and the extent of frustration?

Solution to Exercise 1.2

The arrangement of the three regions is shown in Fig. 1.29. Here, we fix the overlapping region $x = 0.01$, and the width of each rectangle is 1. We change the length of each of the rectangles q from 0.01 to 0.1. The triplet probabilities may be read from Fig. 1.29, these are:

$$P(1, 1, 1) = x$$
$$P(0, 0, 0) = 1 - 3q + 3x - x$$
$$P(1, 0, 0) = P(0, 0, 1) = P(0, 1, 0) = q - x$$

$$P(0, 1, 1) = P(1, 1, 0) = P(1, 0, 1) = 0$$

$$\sum_{x_1, x_2, x_3} P(x_1, x_2, x_3) = x + 1 - 3q + 2x + 3(q - x) = 1 \tag{1.44}$$

Figure 1.30 shows the three SMIs for this system ($x = 0.01$), and changing q. The curves are very similar except for the numerical values which increase from SMI(A) to SMI(A, B), to SMI(A, B, C). Figure 1.31 shows the pair MI and the total triplet MI. Both have a minimum as a function of q.

As usual, the most interesting behavior is that of the conditional MI, CI. In Fig. 1.32 we show $CI(A; B; C)$ as a function of q, and its two components in the equation:

$$CI(A; B; C) = [I(A; B) + I(A; C)] - I(A; (B, C)) \tag{1.45}$$

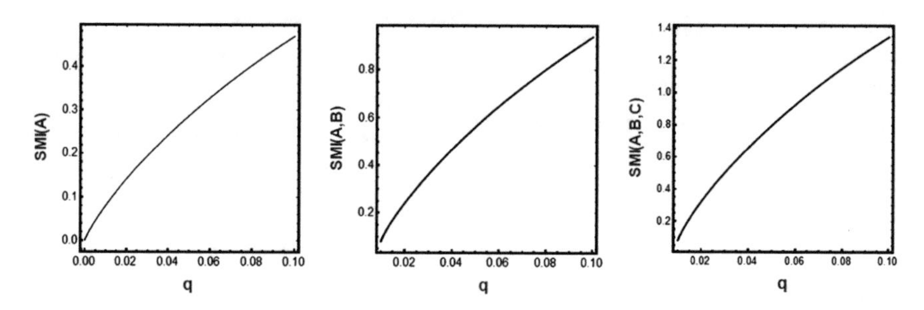

Fig. 1.30 The singlet, pair and the triplet SMI as a function of the distance d

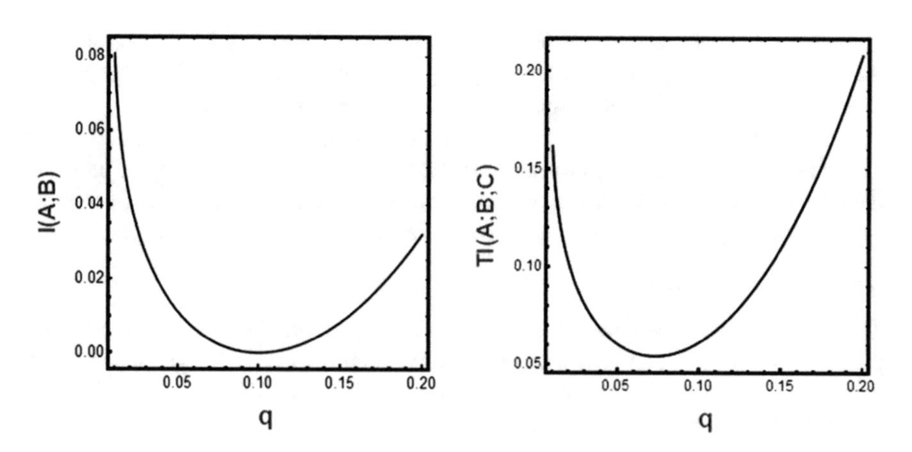

Fig. 1.31 The pair and the total triplet MI as a function of the size of the rectangle q

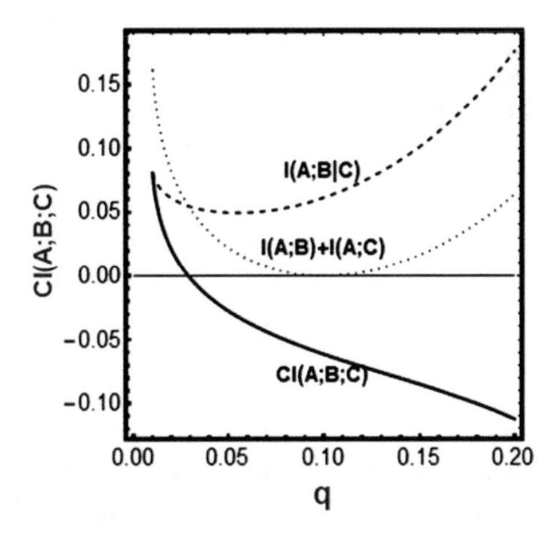

Fig. 1.32 The conditional triplet MI, and the two terms on the right hand side of Eq. (1.57), as a function of the size of the rectangle, q

As can be seen from Fig. 1.32, for small values of q, CI is positive (the sum of the two pair-MI is larger than the conditional MI), but at higher values of q, CI becomes negative.

Again, we ask two questions: Can one express the values of CI as a (positive) area in a Venn diagram, and can one interpret the values of CI in terms of the extent of frustration?

Exercise 1.3 This exercise has a "story" similar to the one mentioned in Sect. 1.5.3. In that story we were told of a crime committed in some region A. A police officer saw the suspect in region B which supports the claim that the suspect was in A. Similarly, another police officer saw the suspect in region C which supports the claim that the suspect was in A at the time of the commission of the crime. However, in the case of Sect. 1.5.3, we saw that the two combined witnesses of the police officers prove that the suspect could not have been in A. This result seems strange. How can each evidence support some claim when the combined evidence does not support the claim?

In the present example, the story is somewhat similar, but with an opposite conclusion.

Suppose that a crime was committed sometime in region A, Fig. 1.33. A suspect with a record of earlier crimes was a suspect in that crime. However, the suspect claimed that he was not in A on the day when the crime was committed. The suspect's friend, F_B told the police that he saw the suspect in region B on that day (see Fig. 1.33. region B overlaps A but the area of x is very small compared to area of either A or B). This evidence does not support the claim that the suspect was in A. Similarly, another friend, F_C told the police that he saw the suspect in region C, again, evidence which does not support the claim that the suspect was in A.

We now have a situation whereby each separate evidence does not support the claim that the suspect was in A at the time of the crime. However, when the judge examined the two evidences, he noticed that the regions B and C intersect, and the

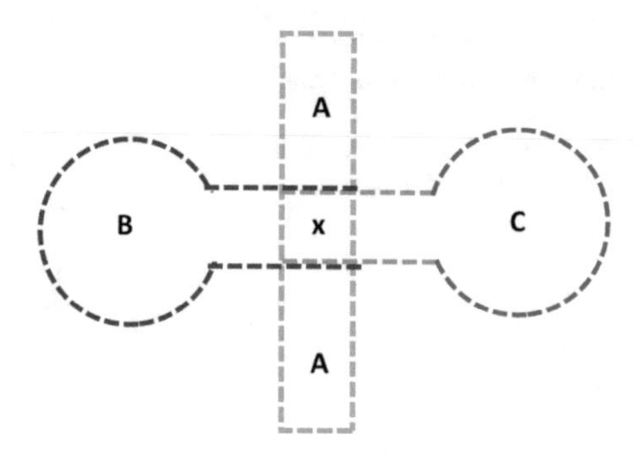

Fig. 1.33 The three regions for Exercise 1.3. The area of the overlapping region between the three rectangles is x

intersection area x is entirely in A, see Fig. 1.33. The judge therefore, concluded that the suspect was *certainly* in A at the time of the commission of the crime.

This story is on the "level" of events and probabilities. Examine the same system at the level of SMI and MI.

Solution to Exercise 1.3

The triplet probabilities may be read from the Fig. 1.33. We assume that the area of A is fixed $a = 0.1$, while each of the regions B and C has an area of$q$, which will be changed between 0.1 to 0.3 (recall that the total area of the board was 1). The three regions A, B, and C intersect at a region with an area of $x = 0.01$.

The triplet probabilities for this arrangement are:

$$
\begin{aligned}
&P(1,1,1) = x \\
&P(0,0,0) = 1 - 2q - a + 2x \\
&P(1,0,0) = a - x \\
&P(0,1,0) = P(0,0,1) = q - x \\
&P(1,1,0) = P(1,0,1) = P(0,1,1) = 0
\end{aligned}
\tag{1.46}
$$

Check that the sum of all $P(x_1, x_2, x_3)$ is one. The reader is urged to examine the various probabilities and the various SMIs (note that $\mathrm{SMI}(A, B) = \mathrm{SMI}(A, C)$, but is different from $\mathrm{SMI}(B, C)$. Both of these have a minimum in which the value of the MI is zero.

Figure 1.34 shows the pair and total triplet MI, which, as expected are everywhere positive. Figure 1.35 shows the conditional triple MI. Interestingly; the values of CI are negative in the entire range of variation of q.

Exercise 1.4 This example is an extension of the two areas discussed in Sect. 1.4.2. We have three consecutive regions A, B, and C, each with the same area $q = 0.1$ (the total area of the board is unity), Fig. 1.36.

The two rectangles A and C move towards B, such that the overlapping area between A and B, and between B and C is x. We change x in two steps: First between

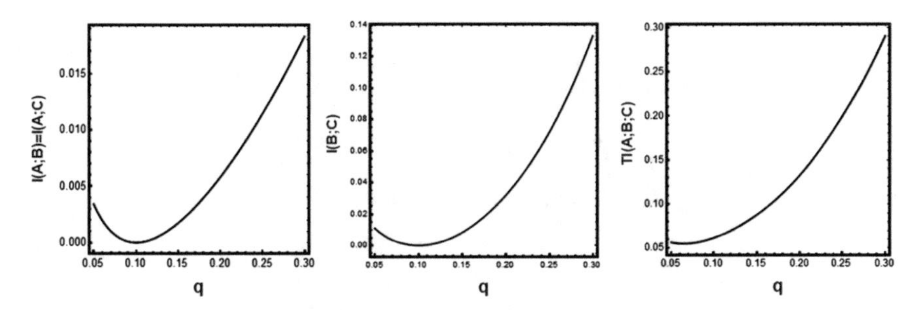

Fig. 1.34 The pair and the total triplet MI as a function of q

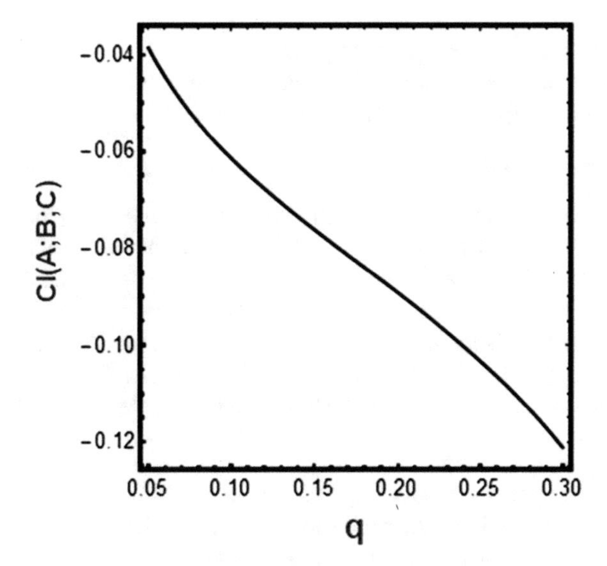

Fig. 1.35 The conditional triplet MI, as a function of q

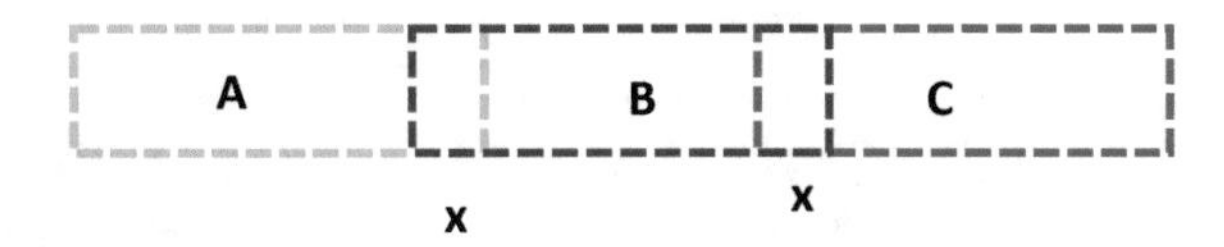

Fig. 1.36 The three regions for Exercise 1.4. The area of the overlapping region between A and B, and between B and C is x

$0 \leq x \leq q/2 = 0.05$, and second, between $0.05 \leq x \leq 0.1$. In the first range of x, A and B overlaps and B and C overlap, but not A and C. In the second range, there is an overlapping area between the three regions. See Fig. 1.37. Examine the SMI and the various MI in this system. Ponder about the possibility of using Venn diagram and the possible interpretation of the negative CI in terms of frustration.

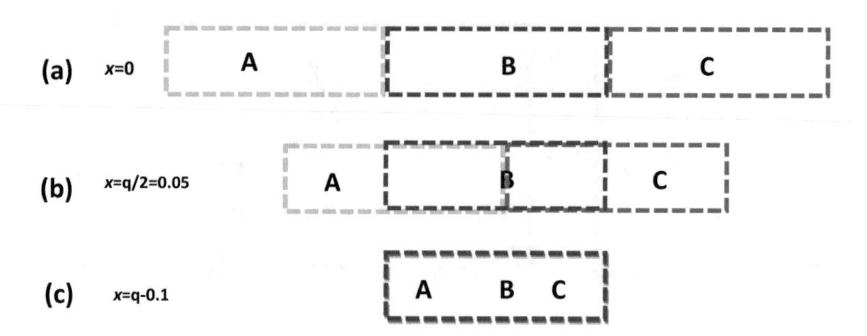

Fig. 1.37 Three stages in the process of changing x in Exercise 1.4

(a) No overlapping between the three pairs.
(b) Overlapping area between A and B, and between B and C, $x = q/2$.
(c) Total overlapping area between A B and C, $x = q$.

Solution of Exercise 1.4

In this example we calculate all the SMI and MI for the two cases; (i) for $0 \le x \le 0.05$ and (ii), for $0.05 \le x \le 0.1$, separately. In Table 1.3 we show the triplet probabilities for these two cases. Note that the probabilities for $x = 0.05$ must be the same in the two columns in Table 1.3, see Fig. 1.37b. In the following we show only the results for the MI. The reader should calculate all the SMI and the MI for the two cases, study them carefully, and try to make sense of the results. Figure 1.38 shows the pair-MI for the two cases, and Fig. 1.39 shows the total triple and conditional MI for the two cases.

Note that at the point $x = 0.05$ (Fig. 1.37b) the results for the two cases must be equal. The relevant limiting values are shown in Table 1.4.

Figure 1.38a shows the pair MI for the case (i), of $0 \le x \le 0.05$. Note that in this case, $I(A; C)$ is constant, independent of x. This is due to the constant values of the SMIs:

Table 1.3 The triplet probabilities

Probability	For $0 \le x \le 0.05$	For $0.05 \le x \le 0.1$
$P(1, 1, 1)$	0	$2x - q$
$P(0, 0, 0)$	$1 - 3q + 2x$	$1 - 3q + 2x$
$P(1, 0, 0)$	$q - x$	$q - x$
$P(0, 1, 0)$	$q - 2x$	0
$P(0, 0, 1)$	$q - x$	$q - x$
$P(1, 1, 0)$	x	$q - x$
$P(0, 1, 1)$	x	$q - x$
$P(1, 0, 1)$	0	0
$\sum P(i, j, k)$	1	1

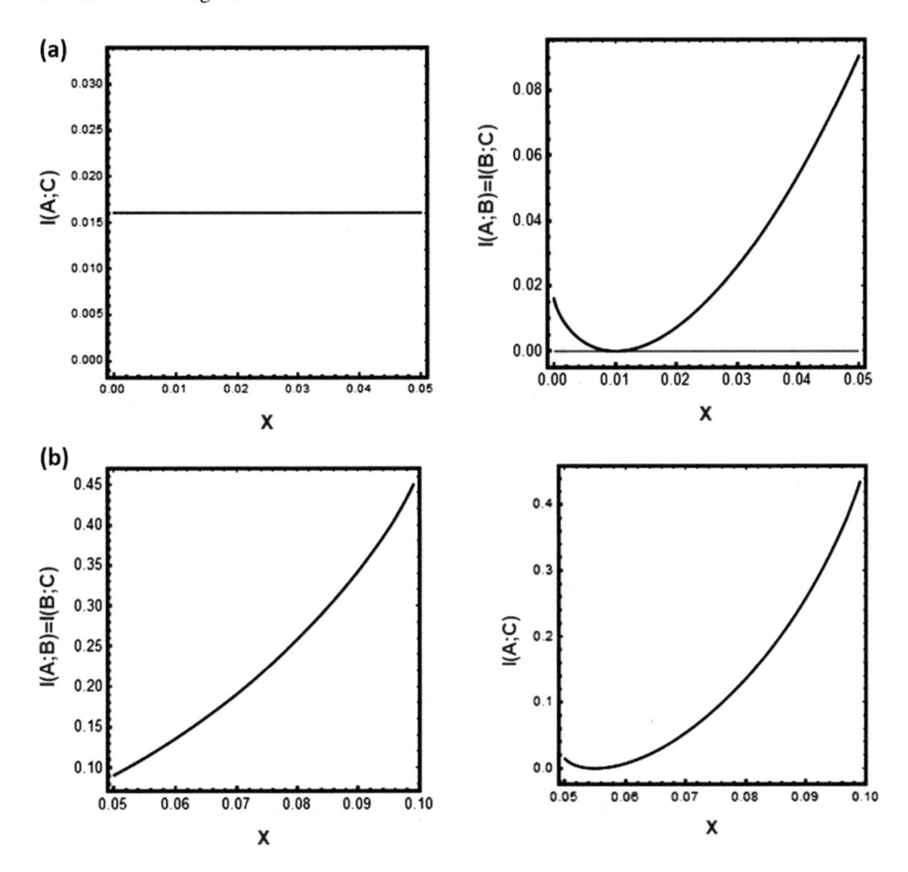

Fig. 1.38 **a** The pair-MI as a function of x, for $0 \leq x \leq 0.05$. **b** The pair-MI as a function of x, for $0.05 \leq x \leq 0.1$

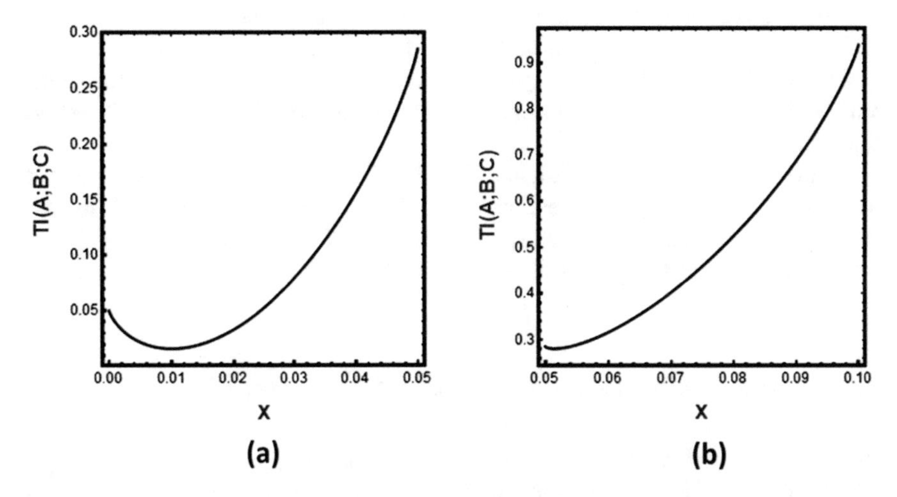

Fig. 1.39 The total triplet MI: **a** for case (i) and **b** for case (ii), as a function of x

Table 1.4 Limiting values at $x = 0.05$

$I(A; B) = I(B; C)$	0.090
$I(A; C)$	0.016
$TI(A; B; C)$	0.285
$CI(A; B; C)$	−0.088

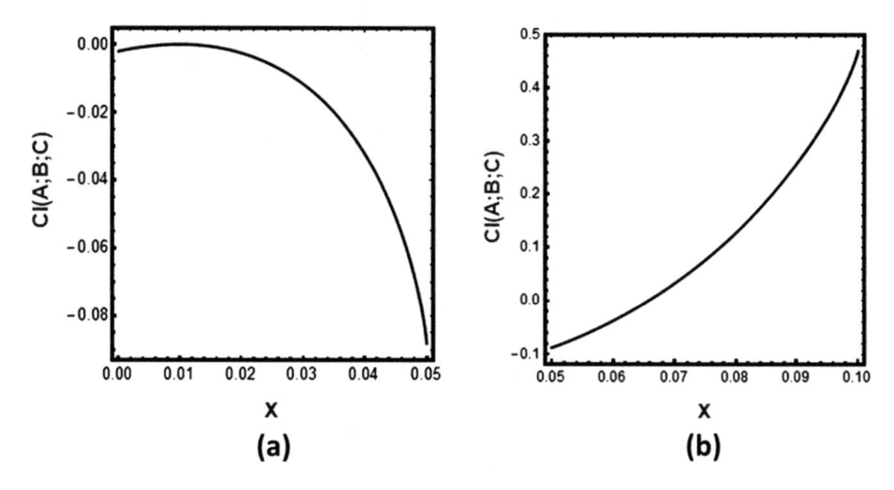

Fig. 1.40 The conditional triplet MI: **a** for case (i) and **b** for case (ii), as a function of x

$$I(A; C) = H(A) + H(C) - H(A, C)$$
$$\approx 2 \times 0.469 - 0.922 \approx 0.016 \tag{1.47}$$

On the other hand, $I(A; B) = I(B; C)$ goes through a minimum at $x = 0.01$ at which point the value of the MI is:$I(A; B) = 0$ (i.e. independence).

Figure 1.38b shows the values of the pair-MI for the case (ii), for $0.05 \le x \le 0.1$.

Figure 1.39a shows the total triple MI for case (i). We see that, as expected all the values are positive and they go through a minimum at $x = 0.01$. Figure 1.39b shows the total triple MI for case (ii).

Figure 1.40 shows the conditional triple MI for the two cases. The behavior of these two cases are quite different. Note that CI are all *negative* in case (i). In case (ii) the value of CI starts with a negative value of -0.088, at $x = 0.05$, and then increase monotonically up to about $CI(A; B; C) \approx 0.47$.

1.6 Levels of Confusion: Information, SMI and Bit

Before we discuss the confusion between the concepts of "information," the measure of information which we call SMI (including the entropy which is a specific case of SMI), and the bit, it is advisable to consider the following examples where the confusion rarely occurs.

Consider the concept of a group, which is a set of objects, people, or even concepts. In mathematics, Group-Theory deals with a particular group or a collection of elements in which we define some operations (say, addition between two numbers). Clearly, there is a difference between the *general* concept of a group, the Group-Theory, and the *specific* group dealt with the in Group-Theory. Example: the general concept of a number, Number-Theory and a specific number say, 5. Clearly, no one would confuse the concept of a number with Number-Theory, and with the specific number 5.

Another example: A nation is a group of people living in some country. We may define measure on either the whole group, or on each of its elements. For example, the size of a nation could be the total number of people, or the total mass of all the people in that nation. We can also define a measure of each individual, its weight, or its length. Then, we can define the units of that measure, such as grams, centimeter, etc.

I am not aware of anyone confusing the concept of a *nation* with the *number* of people in the nation, and the *centimeter.*

Ridiculous as it may sound, exactly this kind of confusion exists between *information* which is a very general concept, *Information Theory*, which is a theory of a specific type of information, the SMI which is a specific measure of a specific type of information, and the *units* with which we measure the SMI. It is unfortunate that in this case, the confusion between these concepts is the rule, rather than the exception.

Information is a very general concept. We all know what information means, though a precise definition of information is elusive. Information may be objective, or subjective. It may be important, interesting or dull. It might be relevant to someone or totally irrelevant to others.

To some information we may assign a size or a measure. "Tomorrow, it will be snowing in New York." This information may be vital to someone living in New York who is planning to engage in an outdoor activity. On the contrary, it could be totally irrelevant to a farmer in a small town in China. To this information we can assign a measure; e.g. the number of words or the number of letters. It can also be its length (in units of centimeter) as written in a specific language with specific fonts.

The SMI is a specific measure defined on a specific type of information. SMI is not information. As we have discussed in Ben-Naim [1], the SMI is defined on a probability distribution. The interpretation of this measure in terms of the amount of information is valid only for this particular type of information. This measure is always an *objective* quantity. See also Sect. 1.7. We can say that this information is contained in, or belongs to the distribution on which it is defined. The units of this measure is the bit (see Sect. 1.1).

Why do people confuse information with SMI, and SMI with the bit? Perhaps, because all these concepts usually include the term "information" in their description. Perhaps, people are not aware of IT, and they do not understand what SMI is.

In fact, the confusion between these three concepts extend beyond these three concepts, and includes other, unrelated concepts: A material object is confused with information ("information is a physical entity," see Landauer quotation below). SMI is confused with entropy (see Sect. 1.3), and the bit in information theory is confused with the bit in binary digits (see Sect. 1.1). Perhaps, the culmination embracing all of this confusion is expressed in the slogan "*It from bit.*" Which is discussed in Ben-Naim [3, 7]. Here we bring one typical quotation from an article by Landauer [15], "*The Physical Nature of Information.*" This article starts with the following two statements:

1. "*Information is not a disembodied abstract entity; it is always tied to a physical representation.*"
2. "*The acceptance of the view, however, that information is physical entity, has been slow.*"

The author seems to consider the two statements: "*Information is…tied to a physical representation*" and "*information is physical entity* as being equivalent. They are not equivalent! Here is why:

The first—"*information is tied to*"—is trivially true. Every abstract concept is tied to some physical entity. This is true for information, as well as for love, hate, beauty, and sociology. Love or hate involves people, beauty is assigned to an object or a person, and sociology deals with a group of people. We can kick a ball, measure the force acting on it, and estimate its velocity. We cannot kick love, beauty or sociology, and we cannot measure their velocity. Why" because they are not physical objects! The same is true with information; have you tried to kick, or to measure the speed of the information about the weather in NY?

The second—"*information is physical entity*"—is obviously not true, as many other abstract concepts are not physical entities.

The confusion between information and SMI, and SMI with entropy abound in the literature. All of these are abstract concepts, and are all "*tied*" to some physical system, but none "*is*" a physical entity. Love is an abstract concept; it is not a physical entity; we cannot say that love is big or small. When love is "tied to a physical entity" we can say that it is great or shallow, but that does not make love a physical entity!

1.7 Information May Be Either Subjective or Objective. The SMI (as Well as Entropy) Is Always an Objective Concept

In Sect. 1.6 we discussed the confusion between the general concept of information, and the SMI. This confusion is quite common. It is also the origin of the popular view that the SMI (as well as entropy) has an element of subjectivity. Before we show why SMI (as well as the entropy) has absolutely no traces of subjectivity we present a few quotations from popular science books. For specific quotations see Ben-Naim [7].

Shannon's measure of Information (SMI) is a *measure* of information. This is clear, but it contains a pitfall. People might ask: If SMI is a measure of information, or a measure of missing information, who has that information, or who is ignorant of that information? This very question implies that a human being is involved in knowing, or not knowing the information, hence, the view that SMI has some degree of subjectivity. In fact, even some serious scientists say this explicitly; if you could *know* the precise microstate in which the system is, then its entropy would be zero. To demonstrate why such view is incorrect, we present here three examples of the SMI.

(a) **A coin hidden in one of eight boxes**

Bob placed a coin in one of eight boxes, Fig. 1.41. Bob tells Linda that the box, in which the coin is, was chosen at random, i.e. with equal probability of 1/8. To eliminate any traces of subjectivity, a random integer between one and eight was chosen and then placed the coin in the box with that number. Linda was also told that there are exactly eight boxes, and that the coin is in one of the boxes. Linda does not know where the coin is, and she has to ask binary questions in order to find out where the coin is.

I tell you, the reader, that the SMI for this game is:

$$\text{SMI(coin in eight boxes)} = \log_2 8 \tag{1.48}$$

I also tell you that this number may be interpreted as a measure of information associated with the distribution $\left(\frac{1}{8}, \frac{1}{8}, \cdots, \frac{1}{8}\right)$ in the following sense: If you know only the distribution, you can find out the missing information on where the coin is, by asking binary questions, and if you are smart enough you are guaranteed to obtain this information with just three questions.

Now, pause and answer the following questions:

(i) Is the SMI for this game a subjective quantity?
(ii) Does the SMI for this game depend on who plays the game?
(iii) Does Bob calculate a different SMI for this game than Linda?

The answer to each of these three questions is No! This seems strange to someone who does not read carefully the description and rules of the game. In this description, we used the word "information" that Bob knows, but Linda doesn't. We also used the word "smart," which might suggest to some that if the person who plays the game is not smart, he or she might calculate a different SMI for this game. All these "words" do not change the fact that the *number*: $\log_2 8 = 3$ is not a subjective number. In the description of the game I told you that Bob placed the coin in one of the boxes, so he must know the information on the location of the coin, while Linda doesn't. However,

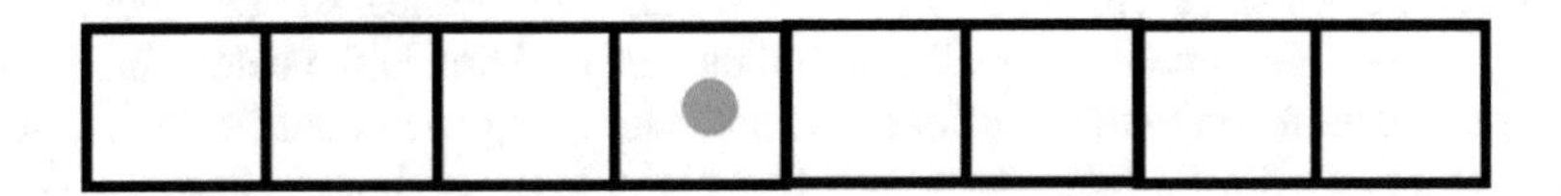

Fig. 1.41 A coin is hidden in one of eight boxes

when I ask you about the SMI that Bob will calculate for this game, the answer is $\log_2 8 = 3$, independently of what Bob knows or doesn't. When Bob plays the game, it means that all he knows is that there are eight equally probable possibilities. With that information he still has to ask three questions.

(b) A dart hit a board divided into eight regions of unequal areas

This game is a little more difficult since it involves a non-uniform distribution.

It is known that a dart was thrown on a board with a unit area. The board is divided into eight regions with areas $p_1, p_2, \cdots, p_8$. It is also known that the dart is in one of those areas and the probabilities of being in one of those regions is proportional to the ratio of the area of that region and the total area of the board (which was chosen as unity). Thus, we know that:

$$\sum p_i = 1$$

And we define the SMI for this distribution as:

$$\text{SMI(dart on eight regions)} = -\sum p_i \log p_i \tag{1.49}$$

The sum is over al $i = 1, 2, \ldots, 8$. Now, we play the same game as before. Bob threw the dart and Linda has to ask binary questions in order to find out where the dart is.

Read questions (i) to (iii) asked in connection with the previous game and answer them. Again, the answers to all those questions is No! Clearly, if the distribution is not uniform the average number of questions one needs to ask in order to obtain the missing information is smaller than $\log_2 8$. This was proven in Chap. 2 of Ben-Naim [1]. However, whatever the distribution is, it determines the value of the SMI as defined in Eq. (1.49), and this value is independent of who plays the game, who knows or does not know where the dart is, and whether or not the game is played at all. The value of the SMI is determined once you are given the distribution, and this number has no element of subjectivity. The game we built upon this distribution, and the identification of specific persons involved in this game are parts of the *interpretation* of the SMI; they do not affect the value of the SMI.

(c) A thermodynamic system with fixed energy *E*, volume *V*, and *N* atoms

This case is similar to the previous two examples I tell you that this system can be in any one of W microstates. In the previous cases W = 8. In our system, a micro-state is a detailed description of the locations and velocities of all particles in the system. In a classical system, there are infinitely many such microstates. But the uncertainty principle imposes some limits on the simultaneous precision in the determination of both the location and velocity of each particle and along each coordinate. Without getting into any details (see Chap. 5 of Ben-Naim [1]), we assume here that W is a finite integer number. For N of the order 10^{23}, W is a huge number (of the order of N^N). Now I tell you that the entropy (which is a special case of an SMI) of this system is given by:

$$\text{Entropy}\left(\begin{array}{c} system\ characterized\ by: \\ E, V, N \end{array}\right) = K \ln W \qquad (1.50)$$

where K is some constant and ln is the natural logarithm.

Is the entropy a subjective quantity? Of course not.

Now, you can construct a similar game on this system. You can imagine that the system's microstates are changing rapidly among all the W possible microstates, which is determined by the specification of the thermodynamic variables E, V, and N.

Suppose that Bob looks inside the system and he observes that the system was in a microstate i at some time t. Bob asked Linda to play the same game as before. She only knows that there are W possible microstates, and she has to find out in which state Bob saw the system at time t, by asking binary questions. Of course, in this particular game there will be many more questions to ask than in the previous games, but the principle is the same, and the questions I want to ask are the same; see (i), (ii), and (iii) above. What are your answers to those questions? What are your answer to the same questions, but instead of entropy we ask about the SMI of the system?

I am confident that the answers you will give to these questions will be the same as in the previous examples. The SMI is determined once the game (or the experiment or the random variable) is given. Similarly, the entropy of the system is determined once the thermodynamic variables are specified. It is absolutely an objective quantity independent of who plays the game, who knows what about the microstates, or on whether the game is played or not.

How did subjectivity sneak into the SMI and the entropy? My answer is that most people, even some physicists confuse the concept of SMI with the concept of information. This confusion leads us to a similar confusion between entropy and information.

As I have explained in Chap. 5 of Ben-Naim [1], entropy, up to a constant is nothing but a special case of an SMI. Given a thermodynamic system characterized by the variable E, V, N at equilibrium, the distribution of locations and velocities is determined. Therefore, the SMI defined on these distributions is also determined. Multiply this SMI by a constant ($k_B \ln 2$) and you get the entropy of the system. It follows that any interpretation one adopts for the SMI is automatically applied to the entropy. This interpretation is referred to as the *informational* interpretation of entropy; meaning interpretation of entropy in terms of SMI. Here are some of the most puzzling conclusions some people draw regarding this interpretation of entropy.

There are people who *do not* accept this interpretation of entropy. Why? Because they know that entropy is an objective quantity. They confuse SMI with information, and since information may be subjective, they *reject* the "information interpretation of entropy." Others who *accept* the "informational interpretation of entropy," but confuse the SMI with information, conclude that since information may be subjective, it also follows that entropy must be subjective.

This situation is puzzling because both groups of scientists are expected to distinguish between an abstract concept of information, and a mathematically well-defined quantity such as SMI or entropy.

Finally, I would like to add a comment of the interpretation of SMI as information, or missing information. Both of these are correct provided that we mean the *amount* of information, or the amount of missing information, respectively. See Ben-Naim [7].

As we have discussed in Ben-Naim [1], the interpretation of SMI in terms of amount of information refers to the information associated with a given distribution. It is not the missing information about the weather in New York, nor the missing information about the author of this book. The amount of information enters only when we interpret the SMI in terms of 20Q game. The larger the missing information or the larger the amount of information one must obtain, hence, a larger number of questions that must be asked to obtain that information.

Thus, for any given 20Q game there is a fixed value of the SMI (depending on the probability distribution). Whether you call it information or missing information depends on "which side of the game you are."

Appendix 1: Venn Diagram for Pair of Events, and for Pair of Random Variables

In Sect. 1.4, we discussed the applicability of Venn diagrams for two (or more) events. Specifically, we discussed the case of two events A and B in a plan where the area of each event represents its probability. The intersection between the two events is a measure of the probability $P(A \cap B)$, i.e. the probability that both A and B occurred. We also showed that there is no monotonic relationship between the area of the intersection between A and B, and the extent of dependence between the two events. The latter is measured by the correlation function defined by:

$$g(A, B) = \frac{P(A, B)}{P(A)P(B)} \tag{1.51}$$

We now proceed from the "level" of pair of events to pair of random variables. We will see why we cannot use the Venn diagram in the same way we use it for a pair of events.

We define the two random variables $\boldsymbol{A}$ and $\boldsymbol{B}$ as follows: We have a board of unit area. We throw a dart on the wall and we know that the dart hit the board. If we denote the event "the dart hit the board" by Ω, then $P(\Omega) = 1$. On this board we draw two rectangles, also denoted by A and B. Each of this rectangle has an area q. Since we chose the total area of the board as unity, it follows that the probability of the dart hitting A is q, and the same probability for B.

Initially, the two rectangles are disjoint, Fig. 1.42a. We move the two rectangles towards each other. In Fig. 1.42b, we see that the intersection area is x. In Fig. 1.42c, we have maximum overlapping area $x = q$.

The case we discussed in Sect. 1.4 was the particular case of the specific pair of events which we can denote as $(A = 1)$ and $(B = 1)$. $(A = 1)$ means that the dart hit the region A, and the event $(A = 1, B = 1)$ means that the dart is in both A and B. However, for the pair random variables, *A, B* we have four pairs of events:

$$\begin{aligned} &(A = 1), (B = 1) \\ &(A = 1), (B = 0) \\ &(A = 0), (B = 1) \\ &(A = 0), (B = 0) \end{aligned} \tag{1.52}$$

The first case was discussed in Sect. 1.4. Here, we can write the correlation function:

$$g(1, 1) = \frac{P(A = 1, B = 1)}{P(A = 1)P(B = 1)} = \frac{x}{q^2} \tag{1.53}$$

(Note the difference in notation from Eq. (1.51) used in Sect. 1.4). As we have seen, $g(1, 1)$ can be both smaller or larger than one (i.e. $\log g(1, 1)$ being negative or positive, see below).

Next, we consider the pair of events $(A = 0)$ and $(B = 0)$ for which the correlation function is:

$$g(1, 0) = \frac{P(A = 1, B = 0)}{P(A = 1)P(B = 0)} = \frac{q - x}{q(1 - q)} \tag{1.54}$$

And the same equation for:

$$g(1, 0) = \frac{P(A = 0, B = 1)}{P(A = 0)P(B = 1)} = \frac{q - x}{q(1 - q)} \tag{1.55}$$

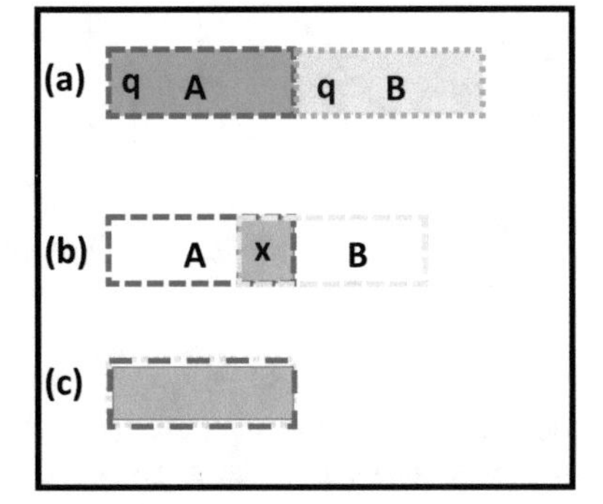

Fig. 1.42 Venn diagram for the two events: (A = 1) and (B = 1). **a** initial x = 0, **b** intermediate x, **c** final x = q. The overlapping area for the intersection (A = 1, B = 1) is shown in grey

The last pair of event is $(A = 0)$, $(B = 0)$, and the corresponding correlation is:

$$g(0,0) = \frac{P(A = 0, B = 0)}{P(A = 0)P(B = 0)} = \frac{(1 - 2q + x)}{(1 - q)^2} \tag{1.56}$$

Note carefully that in Sect. 1.4, we examined the correlation function $g(1, 1)$ as a function of the overlapping area which in this case is x itself. In this particular case, there is a linear relationship between $g(1, 1)$ and x, which we show in Fig. 1.43a. Note that in Fig. 1.43a we drew $g(1, 1)$ as a function of the overlapping area (ov) (while in this particular case it is simply the area x in Fig. 1.43). Note also that $g(1, 1)$ starts at zero. For $x = 0$, it is smaller than unity for $x < q^2 = 0.01$, and its maximum value is $g(1, 1) = 10$ for $x = q = 0.1$, Fig. 1.43a.

For the second pair of events, either $(A = 1)$, $(B = 0)$ or $(A = 0)$, $(B = 1)$, we draw a parametric plot of the correlation as a function of the overlapping area, in Fig. 1.43b. Note that we also have a linear relationship between $g(1, 0) = g(0, 1)$, and the overlapping area, Fig. 1.44. As we can see, the correlation $g(1, 0)$ starts at zero for zero overlapping area, when $x = q$ (see Eq. 1.55). For most of the overlapping areas the values of $g(1, 0)$ is smaller than unity (i.e. negative $\log g(1, 0)$, see below). It is maximum when $x = 0$, $g(1, 0) = 1.11$.

For the case $(A = 0)$, $(B = 0)$ we draw the correlation as a function of the overlapping area, see Figs. 1.43c and 1.45. Note that for smaller overlapping areas, the correlation is less than one, but most of the time the correlation is larger than one (positive correlation).

Thus, we see that for each pair of events we can draw a Venn diagram and also examine the dependence of the correlation function on the extent of overlapping between the two events.

When we study the mutual information, we take an average over all the four correlations (actually the average of $\log_2 g$). In this case, we might have both negative and positive values, but the average of these must be positive. Figure 1.46a shows the three correlations as a function of x. Figure 1.46b shows the pair MI which is the

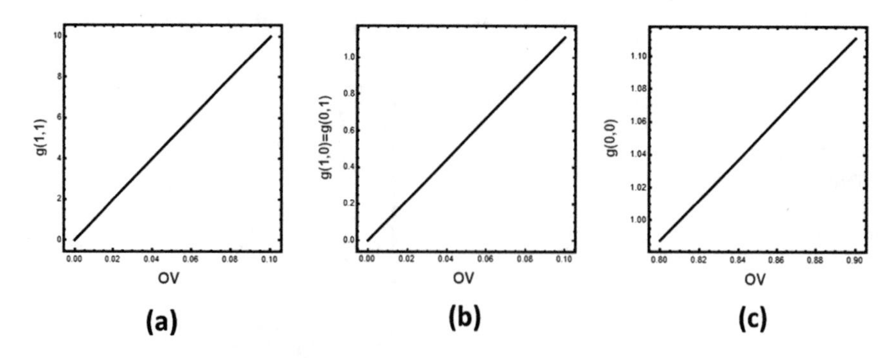

Fig. 1.43 The parametric plot of the correlation functions **a** g(1, 1), **b** g(1, 0) and **c** as a function of the corresponding overlapping area (ov)

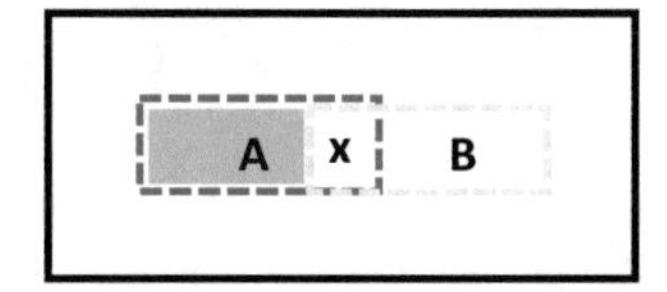

Fig. 1.44 Venn Diagram for the two events: (A = 1) and (B = 0). The overlapping area for the intersection (A = 1, B = 0) is shown in grey

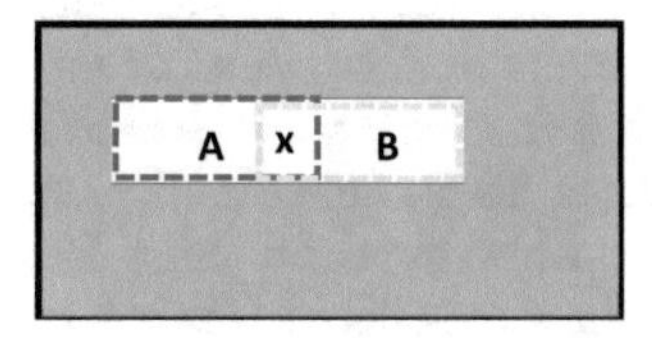

Fig. 1.45 Venn diagram for the two events (A = 0), (B = 0). The overlapping area for (A = 0, B = 0) is shown in grey

average of these functions.

$$I(A; B) = \sum_{ij=0,1} p(i, j) \log[g(i, i)]$$

This quantity is always positive. The same is true when we generalize to any number of random variables, and define the *Total* MI. Unfortunately, this is not always the case for any other definition of a multivariate MI.

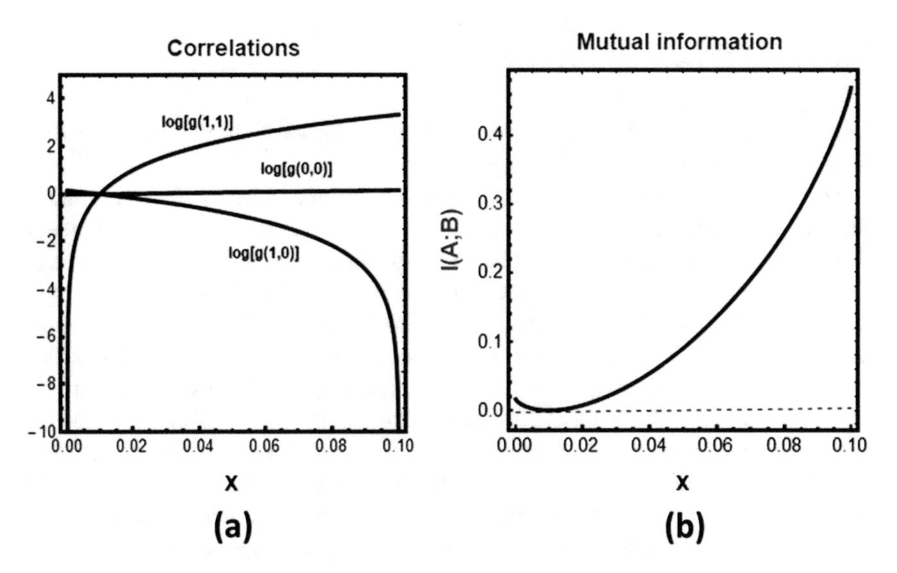

Fig. 1.46 **a** The three correlation functions and **b** the Pair-MI, as a function of x

Appendix 2: The Monty Hall and the Equivalent Three Prisoners' Problem

This is one of the most interesting examples where our intuition fails us in assessing the correct probability. It is also an example where information given on one thing enhances the information one has on another thing.

The Monty Hall Problem got its name from the popular TV game-show called, "Let's Make a Deal," which was hosted by Monty Hall. The player of the game, is given the choice to select one out of three closed doors (or curtains). Behind one of the doors, there is a valuable prize, say a car. Behind the other two doors there is nothing at all. After the player has made the choice, Monty Hall opens one of the remaining two doors, behind which there is nothing. He then suggests to the player: would you like to switch your chosen door to the other unopened door, or stay with your original choice. Would you, the reader, switch doors?

Interestingly, most people believe that it does not matter whether you switch or not. It can be proved mathematically that by switching you increase your chances of winning. In her book, vos Savant discusses at great length the story of the Monty-Hall problem, her answer to the problem, and the reaction of thousands of readers who claimed that her answer to the Monty-Hall problem was wrong. (For a complete solution to the problem based on Bayes' theorem, see Ben-Naim [8, 9].

In the following section we reformulate the problem using a more "dramatic" story about the three prisoners; where your decision can affect your chances of living or dying. The mathematical solution of the problem is identical to the Monty Hall problem.

The Three Prisoners' Problem

This problem is a problem in probability theory. However, in this book, we shall view it as a problem within information theory.

Three prisoners, called A, B, and C, were sentenced to death. A few days before the scheduled execution date, the king decides that he wants to grant freedom to one of the prisoners. The king does not know the prisoners personally and therefore he also does not have any preferred one to be freed. Therefore, he decides to throw a die, the result of which will determine who shall be freed. He makes a special die on the faces of which are inscribed the letters A, A, B, B, C, C. The dice is fair, meaning that the probability of each of the letters, A, B and C, to appear is equal to 1/3. Once the die is thrown and the result is known, the king orders the warden to free the prisoner bearing the name that came up on the face of the die, on the same day that the other two will be executed. He also orders the warden not to reveal to anyone the name of the lucky prisoners until the actual day of the execution.

All the details of the story about the king throwing the die are told to the prisoners.

Prisoner A is pondering his chances of survival. One day before the scheduled execution, he approaches the warden and says: "I know you that you know the results of throwing the die. I also know that you are not allowed to tell me who will be freed tomorrow However, I know for certainty that two of the three prisoners must be executed tomorrow. This means that either B or C will die tomorrow. All I am asking you is that you will tell me which one of these two (B or C) will be executed?

The warren thinks for a while. He concludes that by revealing to prisoner, named A, the identity of one prisoner who will be *executed*, he does not reveal the name of the one who will be *freed*. Hence, by answering prisoner A's question, the warren will not be defying the king's order. Therefore, the warden tells prisoner A that prisoner B is going to be executed the next day.

The new situation is the following: Prisoner A got some new information. He still does not know who will be freed the next day. However, the new *information* giving to him by the warren changes the probability of him being either executed, or freed. The probability problem is the following: Before asking the warden, prisoner A knew that each of the three prisoners have exactly the same probability, i.e. 1/3 of being freed. After getting the answer he has some new *information*. Namely, that B will die. What is the Probability that A will be freed? The "practical" problem may be stated as follows: Supposed the prisoners could have changed their names. Would it be advantageous for A to switch names with C? In probability terms, the question is the following: Prior to asking the warden, A knows that the probability of his survival is 1/3. After receiving the additional *information* (about B), the question is: what is the *conditional probability* of A's survival given the new information?

Interestingly, most people feel that since initially A and C had the same probability of being freed, there should be the same probability for A and C to be freed, even after the warden informed A that B will be executed. However, this intuitive answer is wrong. This problem can be easily solved using Bayes' theorem. We next calculate the probabilities before and after the new *information* is given to A.

Two Solutions to the Three Prisoners' Problem

We start by defining the following three events, and their probabilities:

$$Af = \{\text{Prisoner A is going to be freed}\}$$

$$Pr(Af) = \frac{1}{3}$$

$$Bf = \{\text{Prisoner B is going to be freed}\}$$

$$Pr(Bf) = \frac{1}{3}$$

$$Cf = \{\text{Prisoner C is going to be freed}\}$$

$$Pr(Cf) = \frac{1}{3}$$

We also define the following three events, and their probabilities:

$$Ad = \{\text{Prisoner A is going to die}\}$$
$$Pr(Ad) = \frac{2}{3}$$

$$Bd = \{\text{Prisoner B is going to die}\}$$
$$Pr(Bd) = \frac{2}{3}$$

$$Cd = \{\text{Prisoner C is going to die}\}$$
$$Pr(Cd) = \frac{2}{3}$$

We also denote by $W(\text{B})$ the following event:

$$W(\text{B}) = \{\text{The warden points at B and says that he will be executed}\}$$

The intuitive solution (read carefully)

A asked the question and as a result of the warden's answer, he (A) *knows* that B will die. Therefore, the *conditional probability* that A will be freed, given that B is going to die is written as:

$$\Pr(Af|Bd) = \frac{\Pr(Af \cap Bd)}{\Pr(Bd)}$$
$$\Pr(Af|Bd) = \frac{\Pr(Bd|Af)\Pr(Af)}{\Pr(Bd)} = \frac{1 \times \frac{1}{3}}{\frac{2}{3}} = \frac{1}{2} \tag{1.57}$$

Similarly, we also have:

$$\Pr(Cf|Bd) = \frac{\Pr(Cf \cap Bd)}{\Pr(Bd)} = \frac{\Pr(Bd|Cf)\Pr(Cf)}{\Pr(Bd)} = \frac{1 \times \frac{1}{3}}{\frac{2}{3}} = \frac{1}{2} \tag{1.58}$$

If you accept this "intuitive solution," the conditional probabilities of either events Af or Cf are the same as the initial probabilities, see above. Unfortunately, this solution, though intuitively appealing, is wrong. The reason is that we have used the event *Bd* as the *condition*, in both Eqs. (1.57) and (1.58). Instead we must use the event $W(B)$, as the given condition. Clearly, there is a difference between the "information" included in the event *Bd* and in the event $W(B)$. It is easy to see that the event $W(B)$ is *contained* in Bd. This means, that if one knows that $W(B)$ occurred, then it follows that Bd is true. However, if Bd is true, it does not necessarily follow that $W(B)$ occurred. Thus, the size (or the probability) of the event $W(B)$ is *smaller* than that of Bd. Note that Bd and Bf are mutually exclusive events, i.e., $Bd \cap Bf = \emptyset$.

We already know that $Pr(Bd) = \frac{2}{3}$. By using the theorem of total probability, we can write the event $W(B)$ as:

$$W(B) = W(B) \cap Af\, or\, W(B) \cap Bf\, or\, W(B) \cap Cf \tag{1.59}$$

Hence, the probability of $W(B)$ is:

$$\begin{aligned} Pr(W(B)) &= Pr(W(B)|Af)Pr(Af) + Pr(W(B)|Bf)Pr(Bf) \\ &\quad + Pr(W(B)|Cf)Pr(Cf) \\ &= \frac{1}{2} \times \frac{1}{3} + 0 \times \frac{1}{3} + 1 \times \frac{1}{3} = \frac{1}{2} \end{aligned} \tag{1.60}$$

Equation (1.60) means that the probability that the warden will point at B is the sum of the probabilities of the events on the right hand side of Eq. (1.59).

It should be noted that we have assumed that the warden is indifferent towards A and C; if the event *Af has* occurred, i.e. A was chosen to be freed, then the warden can point at either B or C. This selection is made with probability 1/2. However, if Cf occurred, then the warden does not have a choice, and he must point at B (as a result of A's question). As we noted earlier, the event $W(B)$ is *smaller* than Bd ("smaller" in probability terms means that $W(B) \subset Bd$), This means that $W(B)$ contains more *information* than Bd. Therefore, to obtain the correct solution we must use $W(B)$ instead of Bd in the conditional probabilities. This will be done below.

The second solution based on Bayes' theorem (read carefully)

The correct procedure is as follows: we take $W(B)$ as the "condition" in the conditional probability, and write:

$$\begin{aligned} \Pr(Af|W(B)) &= \frac{\Pr(W(B) \cap Af)}{\Pr(W(B))} \\ &= \frac{\Pr(W(B)|Af)\Pr(Af)}{\Pr(W(B))} \\ &= \frac{\frac{1}{2} \times \frac{1}{3}}{\frac{1}{2} \times \frac{1}{3} + 0 \times \frac{1}{3} + 1 \times \frac{1}{3}} = \frac{1}{3} \end{aligned} \tag{1.61}$$

In words, $\Pr(Af|W(B))$ is the probability that A will be freed given that the warden points at B (remember, this is the new information acquired by A).

Similarly:

$$\Pr(Cf|W(B)) = \frac{\Pr(W(B) \cap Cf)}{\Pr(W(B))}$$

$$= \frac{1 \times \frac{1}{3}}{\frac{1}{2} \times \frac{1}{3} + 0 \times \frac{1}{3} + 1 \times \frac{1}{3}} = \frac{2}{3} \tag{1.62}$$

Now it is clear that by switching names with C, A will *double* his chances of survival. By using the information contained in $W(B)$, we got a different result. The reason is, as we pointed out before, that the amount of information contained in $W(B)$ is larger than in Bd. This fact may be written as:

$$\Pr(W(B)|Bd) = \frac{2}{3} < 1$$
$$\Pr(Bd|W(B)) = 1$$

Note that in the first solution we used the *given* information Bd. This is less than the *available* information to A which is $W(B)$. Therefore, we conclude that the correct solution is (1.61) and (1.62). In other words, by switching names, A will *double* his chances of winning his freedom.

A More General, but Easier to Solve, Problem

As we pointed out above, the solution of the three prisoners runs against our intuition, which tells us that if the probabilities of the two events Af and Cf, were initially equal, these probabilities must be the same after getting the additional information.

There is a simple generalization of the same problem which is easier to understand why the new information given by the warden must change the relative probabilities. This is an interested example where a seemingly more difficult problem is actually easier to understand. Before we discuss the next example try to answer the following questions:

(1) What is the information the prisoner A got?
(2) What is the amount of information the prisoner A got?
(3) Is the amount of information the prisoner A got related to the probability of the event Af?

The more general problem is the following: Instead of 3 prisoners, suppose there are 100 prisoners named by the numbers “1”, “2”, ..., “100”. As before, the king chooses randomly one prisoner to be freed on the day of execution). Initially it is known that only one prisoner is going to get freed. Therefore, the probability that a specific prisoner will be freed is: 1/100. Suppose that prisoner “1” (who knows that out of all the other 99 prisoners, 98 must die) asked the warden, “who, out of the remaining 99 prisoners, will be executed”? The warden points to the following: “2”, “3”, ..., (exclude “7”), ..., “100”, i.e. the warden tells prisoner “1” the “names” of 98 prisoners, except the one named “7”, which will be executed.

Intuitively, it is clear that by switching names with “7”, the prisoner “1” increases his chances of survival from 1/100 to 99/100.

Initially, prisoner "1" knew that he has a 1/100 chance to survive. He also knows that there is probability of 99/100 that one out of the 99 prisoners will survive. By acquiring the information on the 98 prisoners who will be executed, prisoner "1" still has a 1/100 chance of survival, but the chances of survival of "anyone" of the remaining 99 prisoners is now "concentrated" on one prisoner named "7". The latter has now a 99/100 chance of survival. Therefore, "1" should pay any price to switch names with "7".

Note that in this and in the previous example, the prisoner A had some initial *information* (on the probability of being freed), and also had received some additional *information.*

We now answer the three questions we posed above:

The answer to question (1) is simple: The information prisoner A got is "Prisoner B is going to die" The answer to the question (2) is given below. This is the SMI associated with the distribution (1/3, 2/3). The answer to the third question (3), is No. The amount of information is not related to a single probability but to the entire distribution.

Now we pose another question: why is it easier to accept the solution for 100 prisoners than the case with three prisoners? The answer is that in the case of three prisoners A got less information than in the case of 100 prisoners.

Initially, we have three equally likely events, i.e. each prisoner had probability 1/3 of being freed, hence the corresponding SMI is:

$$SMI = -\sum_{i=1}^{3} p_i \log_2 p_i = \log_2 3 \cong 1.585$$

When A asks the warden and receives his answer, there are only two possibilities left, the corresponding SMI is:

$$SMI = -\tfrac{1}{3}\log_2 \tfrac{1}{3} - \tfrac{2}{3}\log_2 \tfrac{2}{3} \equiv 0.918$$

In the more general problem, with 100 prisoners, we start with a uniform probability distribution, i.e. each prisoner had probability of 1/100 of being freed, hence

$$SMI = -\sum_{i=1}^{100} p_i \log_2 p_i = \log_2 100 \cong 6.644$$

and after prisoner "1" receives information from the warden, the SMI reduces to

$$SMI = -\tfrac{1}{100}\log_2 \tfrac{1}{100} - \tfrac{99}{100}\log_2 \tfrac{99}{100} \cong 0.080$$

Note that the reduction in the SMI is much larger in the more general problem, and the reduction in SMI is larger, the larger the number of prisoners. We can use the uncertainty interpretation of the SMI to conclude that in first case there is a reduction in uncertainty from 1.585 to 0.918, but in the second case the reduction is from 6.644 to 0.08. thus in the later example the prisoner got much more information and as a result his uncertainty was reduced to nearly zero.

References

1. Ben-Naim, A. (2017). *Information theory, Part I: An introduction to the fundamental concept.* World Scientific.
2. Shannon, C. E. (1948). A mathematical theory of communication. *Bell System Technical Journal, 27*, 379.
3. Ben-Naim, A. (2015). *Discover probability. How to use it, how to avoid misusing it, and how it affects every aspect of your life.* World Scientific.
4. Shannon, C. E., & Weaver, W. (1949). *The mathematical theory of communication.* The University of Illinois Press.
5. Mackay, D. J. C. (2003). *Information theory.* Cambridge University Press.
6. Brillouin, L. (1962). *Science and information theory.* Academy Press.
7. Ben-Naim, A. (2020). *Entropy: The greatest blunder in the history of science.* Independent publisher, Amazon.
8. Ben-Naim, A. (2008). *A farewell to entropy: Statistical thermodynamics based on information.* World Scientific.
9. Ben-Naim, A. (2015). *Information, entropy, life and the universe. What we know and what we do not know.* World Scientific.
10. Ben-Naim, A. (2018), Time's Arrow (?) *The timeless nature of entropy and the second law of thermodynamics.* Lulu Publishing Services.
11. Ben-Naim, A. (2021). *Best sellers selling confusion on entropy, life and the universe.* Independent publisher, Amazon.
12. Ben-Naim, A. (2012). *Entropy and the second law. Interpretation and misss-interpretationsss.* World Scientific.
13. Ben-Naim, A. (2018). *An informational theoretical approach to the entropy of liquids and solutions, entropy 2018.*
14. Matsuda, H. (2000). *Physical Review E, 62*, 3096.
15. Landauer, R. (1996). The physical nature of information. *Physics Letters A, 217*, 88.
16. Rosenhouse, J. (2009). *The Monty Hall problem.* Oxford University Press.
17. Stone, J. V. (2015). *A tutorial introduction.* Sebtel Press.

Chapter 2
Intermolecular Interactions, Correlations, and Mutual Information

2.1 Introduction

In this chapter we discuss an important example of Mutual Information (MI). In Chap. 5 of Ben-Naim [1], we derived the entropy of an ideal gas. We found that the thermodynamic entropy of an ideal gas is a particular case of an SMI. In this chapter we extend the relationship between entropy and SMI to systems of interacting particle. As a result of this relationship between entropy and SMI, we shall use in this chapter the natural logarithm (i.e. logarithm to be base *e*, rather than the base 2 as is more common in IT). The conversion between SMI and the thermodynamic entropy is achieved by multiplying by the constant ($k_B \ln 2$), where k_B is the Boltzmann constant and ln2 is the natural logarithm of 2. In Sect. 2.2, we start by a brief review of the derivation of the entropy of an ideal gas, i.e. a system of simple particles with negligible intermolecular interactions (by simple particles we mean particles with no internal degrees of freedom, say hard spheres). We will be interested mainly in the MI arising from the *locational* correlations between the particles. Having the expression of the entropy of an ideal gas, we shall proceed to derive the general expression of MI for a system of interacting particles. As we shall soon see whenever we introduce intermolecular interactions, there will be *locational* correlations, and these correlations can be used to define the corresponding MI between the locations of the particles. As we have emphasized in Chap. 1 of this book and in Chap. 5 of Ben-Naim [1], while correlations could either be positive or negative, the MI is always a positive quantity.

Note that the addition of intermolecular interactions is done at a constant temperature. We recall that entropy is related to the distribution of *locations* and *momenta* of all the particles of a system at equilibrium. In this chapter we shall be interested only in changes in the correlations among the locations of the particles. The SMI associated with the momenta (or with the velocities) of the particles is maintained at the equilibrium distributions.

A. Ben-Naim, *Information Theory and Selected Applications*,
https://doi.org/10.1007/978-3-031-21276-5_2

Once we derive the general expression for the MI of interacting particles, we will examine in subsequence sections, some simple examples of systems with interacting particles.

2.2 The General Expression for the SMI of Interacting Particles

In Sect. 5.5 of Ben-Naim [1] we derived the expression for the SMI (as well as the entropy) of an ideal gas. Here, we only briefly describe procedure of obtaining the *entropy function* of an ideal gas from the SMI. This procedure contains four steps:

First, we calculate the locational SMI at equilibrium.
Second, we calculate the momentum SMI at equilibrium.
Third, we add a correction due to the uncertainty principle.
Finally, we add another correction due to the indistinguishability of the particles. The resulting SMI of both locations and momenta leads to the *thermodynamic entropy function*, $S(E, V, N)$, or $S(T, V, N)$. Exactly the same function as was originally derived from the Boltzmann entropy by Sackur [2], and Tetrode [3]. In the following sections we shall be very brief. A more detailed derivation is available in Ben-Naim [1].

2.2.1 First Step: The Locational SMI of a Particle in a 1D Box of Length L

Figure 2.1a shows a particle confined to a one-dimensional (1D) "box" of length *L*. The corresponding *continuous* SMI is:

$$H[f(x)] = -\int f(x) \log f(x) dx \tag{2.1}$$

Note that in Eq. (2.1), the SMI (denote H) is viewed as a *functional* of the function *f(x)*, where $f(x)dx$ is the probability of finding the particle in an interval between x an $x + dx$.

Next, calculate the specific density distribution which maximizes the locational SMI, in (2.1). It is easy to show that the result is (see reference [1]):

$$f_{eq}(x) = \frac{1}{L} \tag{2.2}$$

Since we know that the probability of finding the particle at any interval is 1/L, we may identify the distribution which maximizes the SMI as the *equilibrium* (eq.) distribution. The justification for this is explained in details in Ben-Naim [1, 4]. From

(2.2) in (2.1) we obtain the maximum value of the SMI over all possible locational distribution:

$$H(\text{locations in}1D) = \log L \tag{2.3}$$

Next we admit that the location of the particle cannot be determined with absolute accuracy; there exists a small interval h_x within which we do not care where the particle is. Therefore, we must correct Eq. (2.3) by subtracting $\log h_x$. Thus, we write instead of (2.3), the modified H(locations in 1D) as:

$$H(\text{locations in 1D}) = \log L - \log h_x \tag{2.4}$$

In the last equation we effectively defined H(locations in 1D) for the *finite* number of intervals $n = L/h$. The passage from the infinite to the finite case is shown in Fig. 2.1b. Note that when $h_x \to 0$, H(locations in 1D) diverges to infinity. Here, we do not take the strict mathematical limit, but we stop at h_x which is small enough, but not zero. Note also that the ratio of L and h_x is a pure number. Hence we do not need to specify the units of either L or h_x.

2.2.2 *Second Step: The Velocity SMI of a Particle in a 1D "Box" of Length* L

In the second step we calculate the *probability distribution* that maximizes the (continuous) SMI, subject to two conditions:

$$\int_{-\infty}^{\infty} f(x)dx = 1 \tag{2.5}$$

$$\int_{-\infty}^{\infty} x^2 f(x)dx = \sigma^2 = constant \tag{2.6}$$

In his original paper, Shannon [5] proved that the function $f(x)$ which maximize the SMI in (2.1), subject to the two conditions (2.5) and (2.6), is the Normal distribution, i.e.:

$$f_{eq}(x) = \frac{\exp[-x^2/2\sigma^2]}{\sqrt{2\pi\sigma^2}} \tag{2.7}$$

Note again that we use the subscript *eq.* for equilibrium. Applying this result to a classical particle having average kinetic energy $\frac{m<v_x^2>}{2}$, and using the relationship between the standard deviation σ^2 and the temperature of the system:

$$\sigma^2 = \frac{k_B T}{m}, \tag{2.8}$$

we obtain the *equilibrium velocity distribution* of one particle in a 1D system. This is shown in Fig. 2.2:

$$f_{eq}(v_x) = \sqrt{\frac{m}{2\pi k_B T}} \exp\left[\frac{-mv_x^2}{2k_B T}\right] \tag{2.9}$$

Here, k_B is the Boltzmann constant, m is the mass of the particle, and T the absolute temperature.

The value of the (continuous) SMI for this probability density is:

$$H_{\max}(\text{velocity in} 1D) = \frac{1}{2}\log(2\pi e k_B T/m) \tag{2.10}$$

Recall that in calculating the distribution of locations Eq. (2.2) and of velocities Eq. (2.9), we obtained is the distribution that maximizes the SMI. As it is well known these distributions are the same as the *equilibrium distribution.* Therefore, we add the subscript *eq.* in Eqs. (2.2) and (2.9). From the velocity distribution we can get the momentum distribution in 1D, by transforming from $v_x \rightarrow p_x = mv_x$, to get:

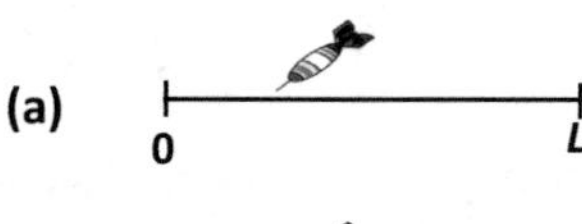

Fig. 2.1 **a** A dart hits a one-dimensional "box" of length L. **b** Passage from the infinite to the discrete number of states

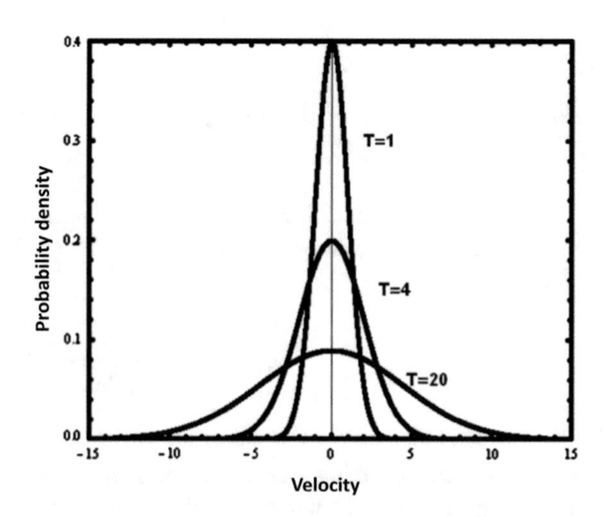

Fig. 2.2 Equilibrium probability density of velocities in a one dimensional system at different temperatures

$$f_{eq}(p_x) = \frac{1}{\sqrt{2\pi m k_B T}} \exp\left[\frac{-p_x^2}{2mk_B T}\right] \tag{2.11}$$

From this distribution we obtain the maximal value of the SMI over all possible densities that fulfill the two conditions (2.5) and (2.6):

$$H_{max}(\text{momentum in} 1D) = \frac{1}{2}\log(2\pi e m k_B T) \tag{2.12}$$

Again, as we did in the previous section, we recognize the fact that there is a limit to the accuracy within which we can determine the velocity (or the momentum) of the particle Therefore, we correct the expression in (2.12) by subtracting log h_p where h_p is a small, but finite interval. The resulting SMI is:

$$H_{max}(\text{momentum in } 1D) = \frac{1}{2}\log(2\pi e m k_B T) - \log h_p \tag{2.13}$$

As we noted before we do not have to specify the units of $h_p(masslength/time)$ and of $\sqrt{mk_B T}$, separately, since the whole expression under the logarithm of the combined two terms in Eq. (2.13) is a pure number.

2.2.3 *Third Step: Combining the SMI for the Location and Momentum of a Particle in a 1D System. Addition of Correction Due to Uncertainty*

If the location and the momentum (or velocity) of the particles were independent events, then the joint SMI of location and momentum would be the sum of the two SMIs in Eqs. (2.4) and (2.12). Therefore, for this case we write:

$$\begin{aligned} H_{max}(l\text{ocation and momentum}) &= H_{max}(\text{location}) + H_{max}(\text{momentum}) \\ &= \log\left[\frac{L\sqrt{2\pi e m k_B T}}{h_x h_p}\right] \end{aligned} \tag{2.14}$$

It should be noted that in the very writing of Eq. (2.14), the assumption is made that the location and the momentum of the particle are independent. However, quantum mechanics imposes restriction on the accuracy in determining both the location x and the corresponding momentum p_x. Originally, the two quantities h_x and h_p that we defined above, were introduced because we did not care to determine the location and the momentum with an accuracy better than h_x and h_p, respectively. Now, we must acknowledge that quantum mechanics imposes upon us the uncertainty condition, about the accuracy with which we can determine *simultaneously* both the location and the corresponding momentum of a particle. This means that in Eq. (2.14), h_x and h_p cannot both be arbitrarily small; their product must be of the order of Planck constant $h = 6.626 \times 10^{-34} Js$. Therefore, we introduce a new parameter h, which replaces the product:

$$h_x h_p \approx h \tag{2.15}$$

Accordingly, we modify Eq. (2.14) to:

$$H_{max}(\text{location and momentum}) = \log\left[\frac{L\sqrt{2\pi emk_B T}}{h}\right] \tag{2.16}$$

2.2.4 *The SMI of One Particle in a Box of Volume* V

Figure 2.3 shows one simple particle in a cubic box of volume V.

To proceed from the 1D to the 3D system, we assume that the locations of the particle along the three axes x, y and z are independent. With this assumption, we can write the SMI of the location of the particle in a cube of edges L, as a *sum* of the SMI along x, y, and z, i.e.

$$H(\text{location in 3D}) = 3H_{\max}(\text{location in 1D}) \tag{2.17}$$

We can do the same for the momentum of the particle if we assume that the momentum (or the velocity) along the three axes x, y and z are independent. Hence, we can write the SMI of the momentum as:

$$H_{\max}(\text{momentum in 3D}) = 3H_{\max}(\text{momentum in 1D}) \tag{2.18}$$

We can now combine the SMI of the locations and momenta of one particle in a box of volume V, taking into account the uncertainty principle, to obtain the result:

$$H_{\max}(\text{location and momentum in 3D}) = 3\log\left[\frac{L\sqrt{2\pi emk_B T}}{h}\right] \tag{2.19}$$

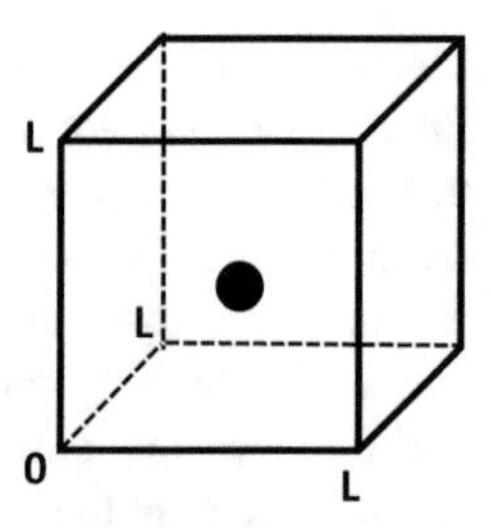

Fig. 2.3 A single particle in a box of volume $V = L^3$

2.2.5 *The Forth Step: The SMI of Locations and Momenta of* N *Independent Particles in a Box of Volume* V. *Adding a Correction Due to Indistinguishability of the Particles*

The final step is to proceed from a single particle in a box, to N independent particles in a box of volume V, Fig. 2.4.

We say that we know the *microstate* of the particle, when we know the location (x, y, z), and the momentum $\left(p_x, p_y, p_z\right)$ of one particle within the box. For a system of N *independent* particles in a box, we can write the SMI of the system as N times the SMI of one particle, i.e., we write:

$$\text{SMI}(N \text{ independent particles}) = N \times \text{SMI}(\text{one particle}) \tag{2.20}$$

This is the SMI for N independent particles. In reality, there could be correlation among the microstates of all the particles. We shall mention here correlations due to the *indistinguishability* of the particles, and correlations is due to *intermolecular interactions* among all the particles. We shall discuss these two sources of correlation separately. Recall that the microstate of a single particle includes the location and the momentum of that particle. Let us focus on the location of one particle in a box of volume V. We write the locational SMI as:

$$H_{\max}(\text{location}) = \log V \tag{2.21}$$

For N independent particles, we write the locational SMI as:

$$H_{\max}(\text{locations of N particles}) = \sum_{i=1}^{N} H_{\max}(\text{one particle}) \tag{2.22}$$

Since in reality, the particles are indistinguishable, we must correct Eq. (2.22). We define the *mutual information* corresponding to the correlation between the particles as:

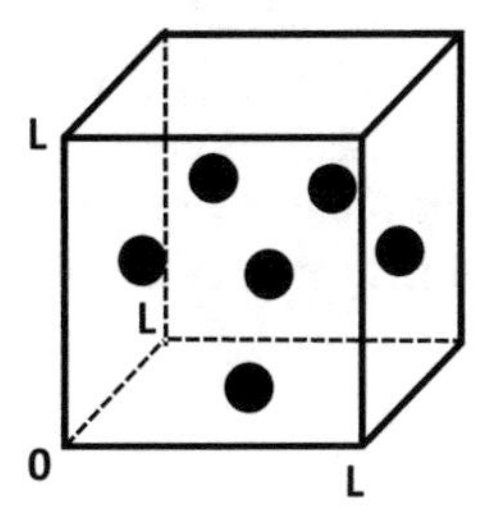

Fig. 2.4 $N = 6$ particles in a box of volume $V = L^3$

$$I(1; 2; \ldots ; N) = \ln N! \tag{2.23}$$

Hence, instead of (2.22), for the SMI of N indistinguishable particles, will write:

$$H(Nparticles) = \sum_{i=1}^{N} H(oneparticle) - \ln N! \tag{2.24}$$

A detailed justification for introducing $\ln N!$ as a correction due to indistinguishability of the particle is discussed in Sect. 5.2 of Ben-Naim [1]. Here we write the final result for the SMI of N *indistinguishable* (but non-interacting) particles as:

$$H(N\ indistinguishable\ \text{particles}) = N\log V\left(\frac{2\pi mek_B T}{h^2}\right)^{3/2} - \log N! \tag{2.25}$$

We are interested in systems with very large N, for which we can use the Stirling approximation for $\log N!$ (note that we use here the natural logarithm) in the form:

$$\log N! \approx N\log N - N \tag{2.26}$$

With this approximation, we obtain the final result for the SMI of N *indistinguishable* particles in a box of volume V, and temperature T:

$$H(1, 2, \ldots, N) = N\log\left[\frac{V}{N}\left(\frac{2\pi mk_B T}{h^2}\right)^{3/2}\right] + \frac{5}{2}N \tag{2.27}$$

This is almost the *entropy function* of a system of N *indistinguishable* particles in a box of volume V, at temperature T. To obtain the entropy we need to multiply by a constant (k_B, if we used the natural logarithm, or $k_B\log_e 2$ if we used the logarithm is to the base 2). Once we do that we get the same equation was derived by Sackur [2] and by Tetrode [3] in 1912, by using the Boltzmann definition of entropy, and applied it to an ideal gas. At this point we have shown that the entropy of an ideal gas as defined by Boltzmann, is identical with the entropy defined in Ben-Naim [1], based on SMI.

In Eq. (2.27) we have a function of the thermodynamic variables (T, V, N). We can easily convert this expression into the *entropy function* $S(E, V, N)$, by using the relationship between the total kinetic energy of the system, and the total kinetic energy of all the particles

$$E = N\frac{mv^2}{2} = \frac{3}{2}Nk_B T \tag{2.28}$$

The explicit *entropy function* of an ideal gas is:

$$S(E, V, N) = Nk_B \ln\left[\frac{V}{N}\left(\frac{E}{N}\right)^{3/2}\right] + \frac{3}{2}k_B N\left[\frac{5}{3} + \ln\left(\frac{4\pi m}{3h^2}\right)\right] \quad (2.29)$$

We refer to this equation as the *entropy function.* It should be noted that it is only for *isolated* systems (i.e. systems with fixed values of (E, V, N) for which the entropy formulation of the Second Law applies, see Ben-Naim [6].

2.2.6 The Entropy of a System of Interacting Particles. Correlations Due to Intermolecular Interactions

In this section we derive the most general relationship between the SMI (or the entropy) of a system of interacting particles, and the corresponding mutual information (MI). Later on in this chapter we shall apply this general result to some specific cases. The implication of this result is very important in interpreting the concept of entropy in terms of SMI. In other words, the "*informational interpretation*" of entropy is effectively extended for all systems of interacting particles at equilibrium.

We start with some basic concepts from classical statistical mechanics [7]. The classical canonical partition function (PF) of a system characterized by the variable T, V, N, *is*:

$$Q(T, V, N) = \frac{Z_N}{N!\Lambda^{3N}} \quad (2.30)$$

where Λ^3 is called the momentum partition function (or the de Broglie wavelength), and Z_N is the configurational PF of the system"

$$Z_N = \int \cdots \int dR^N \exp[-\beta U_N(R^N)] \quad (2.31)$$

Here, $U_N(R^N)$ is the total interaction energy among the N particles at a configuration $R^N = R_1, \cdots, R_N$. Statistical thermodynamics provides the probability density for finding the particles at a specific configuration $R^N = R_1, \cdots, R_N$, which is:

$$P(R^N) = \frac{\exp[-\beta U_N(R^N)]}{Z_N} \quad (2.32)$$

where $\beta = (k_B T)^{-1}$ and T the absolute temperature. In the following we chose $k_B = 1$. This will facilitate the connection between the entropy-change and the change in the SMI. When there are no intermolecular interactions (ideal gas), the

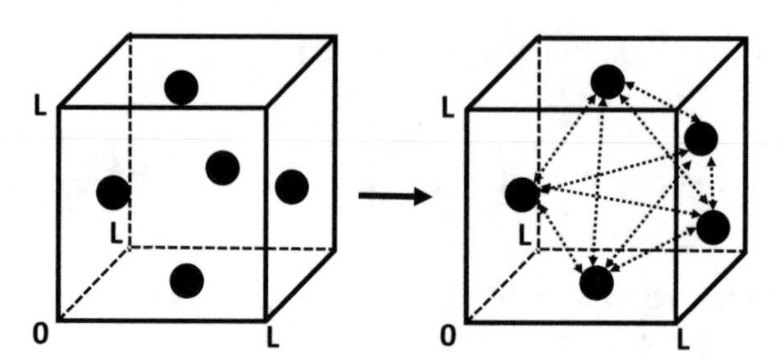

Fig. 2.5 Passage from a system of non-interacting to interacting particles. The pair-interactions are schematically shown by the double arrows connecting each pair of particles. For $N = 5$ we have 10 pairs of interacting particles

configurational PF is $Z_N = V^N$, and the corresponding partition function is reduced to:

$$Q^{ig}(T, V, N) = \frac{V^N}{N!\Lambda^{3N}} \tag{2.33}$$

Next we define the change in the *Helmholtz energy* (A) due to the interactions as:

$$\Delta A = A - A^{ig} = -T\ln\frac{Q(T, V, N)}{Q^{ig}(T, V, N)} = -T\ln\frac{Z_N}{V^N} \tag{2.34}$$

This change in Helmholtz energy corresponds to the process of "turning-on" the interaction among all the particles at constant (T, V, N), Fig. 2.5.

The corresponding change in the entropy is:

$$\begin{aligned}\Delta S &= -\frac{\partial \Delta A}{\partial T} = \ln\frac{Z_N}{V^N} + T\frac{1}{Z_N}\frac{\partial Z_N}{\partial T} \\ &= \ln Z_N - N\ln\mathrm{V} + \frac{1}{T}\int dR^N P\left(R^N\right)U_N\left(R^N\right)\end{aligned} \tag{2.35}$$

We now substitute $U_N\left(R^N\right)$ from (2.36) into (2.35) to obtain the expression for the change in entropy corresponding to "turning on" the interactions:

$$\Delta S = -N\ln V - \int P\left(R^N\right)\ln P\left(R^N\right)dR^N \tag{2.36}$$

Clearly, that the second term on the right hand side of (2.36) has the form of a SMI. It is easy to write also the first term on the right hand side of (2.36) as SMI of ideal gas, for which $U_N\left(R^N\right) = 0$, and the corresponding probability distribution is:

$$P^{ig}\left(R^N\right) = P(R_1)P(R_2)\cdots P(R_N) = \left(\frac{1}{V}\right)^N \tag{2.37}$$

Hence, we can write for the entropy-change:

$$\Delta S = \ln P^{ig}(R^N) - \int P(R^N) \ln P(R^N) dR^N \tag{2.38}$$

Since $P^{ig}(R^N) = (1/V)^N$ is independent of R^N, we can rewrite $\ln P^{ig}(R^N)$ as:

$$\ln P^{ig}(R^N) = \int P^{ig}(R^N) \ln P^{ig}(R^N) dR^N \tag{2.39}$$

Hence, Eq. (2.38) is rewritten as:

$$\Delta S = -\int P(R^N) \ln\left[\frac{P(R^N)}{P^{ig}(R^N)}\right] dR^N = -\int P(R^N) \ln\left[\frac{P(R^N)}{\prod_{i=1}^{N} P(R_1)}\right] dR^N \tag{2.40}$$

We can now recognize the last expression, on the right hand side of (2.40), as the *mutual information* (MI) [1]. To see that, we define the *total correlation* function among the N particles as:

$$g(1, 2, \cdots, N) = \frac{P(R^N)}{\prod_{i=1}^{N} P(R_i)} \tag{2.41}$$

Using (2.41) in (2.40) we get the final form of the entropy-change:

$$\Delta S = -\int P(R^N) \ln g(R^N) dR^N = -I(1; 2; \cdots ; N) \tag{2.42}$$

Thus, except for the Boltzmann constant and change in the base of the logarithm, the entropy-change ΔS is equal to the negative *mutual information* $I(1; 2; \cdots ; N)$, between the locations of the N particles. Since the mutual information is always positive (see Sect. 3.3 in Ben-Naim [1]), the change in entropy in (2.42) must always be *negative.* Note that in Eq. (2.42) we have defined the total MI between the N particles. We can conclude that for any type of intermolecular interactions, whenever we "turn-off" the interactions, the entropy of the system will always *decrease.*

To summarize, we started with an ideal gas for which we know the *entropy function* (either from Boltzmann's definition of from Ben-Naim's definition [1]). We next "turn-on" the intermolecular interactions (here at constant T, V, N). These interactions creates *correlations* among the locations of all the particles. (Note that T constant means no change in the velocity-distribution). These correlations produce *mutual information* MI among the locations of all particles. Note that the correlation function as defined in Eq. (2.41) may either be greater or smaller than one. Therefore, $\ln g(1, 2, \ldots N)$ may either be positive or negative. However, the *average* value of $\ln g(1, 2, \ldots N)$ must be positive. Equation (2.42) is very general. It is valid for any kind of interactions among the particles. In the following sections we shall discuss the case of pairwise interaction energy only. We shall examine the cases of positive and negative correlations, but the MI is always positive.

2.3 The SMI of a System of Interacting Particles in Pairs Only

In this section we consider a special case of a system of interacting particles. We start with an ideal gas—i.e. system for which we can neglect all intermolecular interactions. Strictly speaking, such a system does not exist. However, if the gas is very dilute such that the average intermolecular distance is very large the system behaves as if there are no interactions among the particle.

Next, we increase the density of the particles. At first we shall find that pair-interactions affect the thermodynamics of the system. Increasing further the density, triplets, quadruplets, and so on interactions, will also affect the behavior of the system. In the following we provide a very brief description of the *first order deviation* from ideal gas; systems for which one must take into account pair-interactions but neglect triplet and higher order interactions. The reader who is not interested in the details of the derivation can go directly to the result in Eq. (2.51) and the following analysis of the MI.

We start with the general configurational PF of the system, Eq. (2.31) which we rewrite in the form:

$$Z_N = \int dR^N \prod_{i<j} \exp[-\beta U_{ij}] \tag{2.43}$$

where U_{ij} is the pair potential between particles i and j. It is assumed that the total potential energy is pairwise additive.

Define the so-called Mayer *f-function,* by:

$$f_{ij} = \exp(-\beta U_{ij}) - 1 \tag{2.44}$$

We can rewrite Z_N as:

$$Z_N = \int dR^N \prod_{i<j} (f_{ij} + 1) = \int dR^N \left[1 + \sum_{i<j} f_{ij} + \sum f_{ij} f_{jk} + \cdots \right] \tag{2.45}$$

Neglecting all terms beyond the first sum, we obtain:

$$Z_N = V^N + \frac{N(N-1)}{2} \int f_{12} dR^N = V^N + \frac{N(N-1)}{2} V^{N-2} \int f_{12} dR_1 dR_2 \tag{2.46}$$

The factor $N(N-1)/2$ is the *number of pairs* in a system of N particles.

We now identify the *second virial coefficient* as:

$$B_2(T) = \frac{-1}{2V} \int_V \int_V f_{12} dR_1 dR_2 \tag{2.47}$$

With this identification, we rewrite Z_N as:

$$Z_N = V^N - N(N-1)V^{N-1}B_2(T) = V^N\left[1 - \frac{N(N-1)}{V}B_2(T)\right] \tag{2.48}$$

The corresponding Helmholtz energy change is:

$$\Delta A = A - A^{ig} = -T\ln\frac{Z_N}{V^N} = -T\ln\left[1 - \frac{N(N-1)}{2V^2}\iint f_{12}(R_1, R_2)dR_1dR_2\right] \tag{2.49}$$

This is the change in the Helmholtz energy for "turning-on" the interaction. See Eq. (2.34).

Since we have taken the low density limit, we can rewrite (2.49) as:

$$\Delta A \approx T\frac{N(N-1)}{2V^2}\int f_{12}(R_1, R_2)dR_1dR_2 \tag{2.50}$$

Note that for large N, $\frac{N(N-1)}{V^2} \approx \rho^2$, hence we can use the approximation $\ln(1-\rho^2 B) \approx -\rho^2 B$, where B is the integral in (2.49).

In this limit the change in entropy for the process of "turning on" the interactions is:

$$\Delta S \approx -\frac{N(N-1)}{2V^2} \times \left[\int f_{12}(R_1, R_2)dR_1dR_2 + T\int \frac{\partial f_{12}}{\partial T}dR_1dR_2\right] \tag{2.51}$$

We now use the following limiting behavior of the pair distribution and the pair correlation function. For details, see Ben-Naim [7]:

$$P(R_1, R_2) = \frac{g(R_1, R_2)}{V^2} \tag{2.52}$$

$$g(R_1, R_2) = \exp[-\beta U_{12}] \tag{2.53}$$

$P(R_1, R_2)dR_1dR_2$ is the probability of finding one particle at R_1 and a second particle at R_2.

The correlation function $g(R_1, R_2)$ is a measure of the deviation of the pair distribution $P(R_1, R_2)$ and the product of the two singlet distributions $P(R_1) = P(R_2) = \frac{1}{V}$.

The corresponding entropy-change is:

$$\Delta S = -\frac{N(N-1)}{2} \int P(R_1, R_2) \ln g(R_1, R_2) dR_1 dR_2 \tag{2.54}$$

We see that, except for the base of the logarithm, the integral in Eq. (2.54) is the mutual information (MI) for a specific pair of particles (1 and 2). In a system of N particles there are altogether $\frac{N(N-1)}{2}$ pairs of particles. Therefore, the entropy-change, up to a constant is the mutual interaction between all the pairs of particles in the system, i.e.:

$$\Delta S = -\frac{N(N-1)}{2} I(1; 2) \tag{2.55}$$

Since the mutual information is always positive, ΔS will always be negative. We now examine a particular case of a pair potential. We shall see how both positive and negative correlations lead always to a positive MI, i.e. a negative change in entropy in Eq. (2.55).

The general form of a pair potential is shown in Fig. 2.6. Without getting into details we see that the function $U(r)$, referred to as the pair potential has two regions:

1. Between $0 \leq r \lesssim \sigma$, the function is positive, and the slop is negative.
2. For $r \gtrsim \sigma$, the function is negative.

Note that the function rises steeply for distances shorter than σ. This is the reason for referring to σ [i.e. the value of $U(R = \sigma) = 0$] as the "effective diameter" of the particles. For hard spheres the "effective" diameter is the actual diameter of the spheres.

Statistical mechanics establishes a connection between the pair potential and the pair correlation function, See Ben-Naim [7]. For our case, the corresponding correlation function, Eq. (2.53) has also two regions:

1. For $0 \leq r < \sigma$, $g(r)$ is smaller than one, i.e. negative correlation.

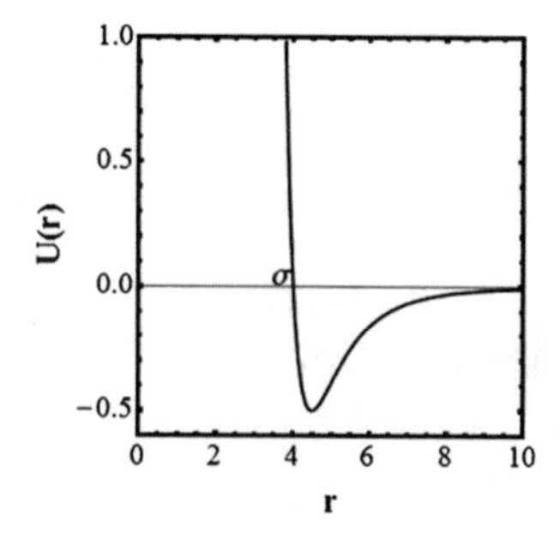

Fig. 2.6 The general form of the pair potential between two particles

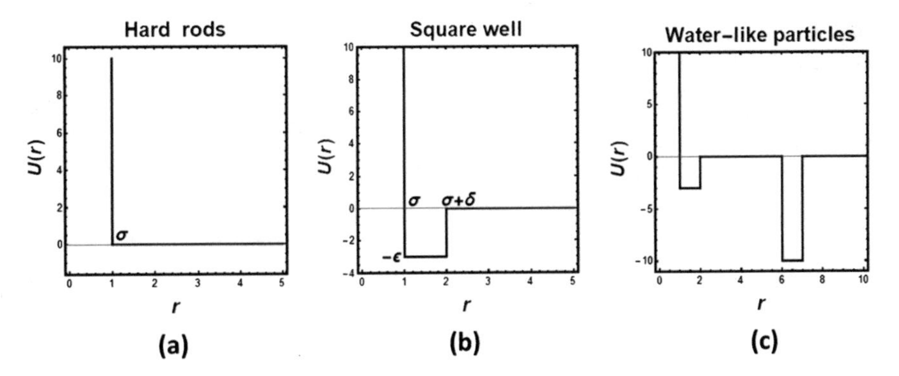

Fig. 2.7 **a** Hard spheres, **b** square well potential between two particles and **c** pair interaction between water-like particles

2. For $r \gtrsim \sigma$, $g(r)$ is larger than one, i.e. positive correlation, Fig. 2.6.

Instead of the pair potential shown in Fig. 2.6 we shall use two approximate functions known as "hard-sphere," Fig. 2.7a and "square-well" potential, Fig. 2.7b.

The square-well function is defined as:

$$U_{sq}(r) = \begin{cases} r \leq \sigma & \infty \\ \sigma \leq r \leq \sigma + \delta & -\varepsilon \\ r \gtrsim \sigma + \delta & 0 \end{cases} \tag{2.56}$$

Thus, for $r \leq \sigma$ the function has a value of ∞. This means infinite repulsive forces, and therefore σ in this case is referred to as the hard-sphere diameter. Two particles cannot come to a distance closer than $r = \sigma$. For the case $\delta = 0$and $\varepsilon = 0$, we get the hard-sphere pair potential.

The region between $\sigma < r < \sigma + \delta$ is supposed to represent the region of negative value in the function $U(r)$ in Fig. 2.6. The value $-\varepsilon$ is referred to as the strength of the pair interaction and δ is the range of the interaction. The corresponding pair correlation function is given by Eq. (2.53).

Note that in the range $0 \leq r \leq \sigma$ the correlation is negative, i.e.

$$\ln g(r) < 0 for\, 0 \leq r \leq \sigma \tag{2.57}$$

In terms of conditional probabilities, we have

$$P(R_2|R_1) = g(R_1, R_2)P(R_2) \tag{2.58}$$

A negative correlation means that the conditional probability of finding a particle at R_2 given another particle at R_1 is smaller than the singlet probability $P(R_2)$. In the case of $r = |R_2 - R_1| \leq \sigma$, Eq. (2.58) is reduced to:

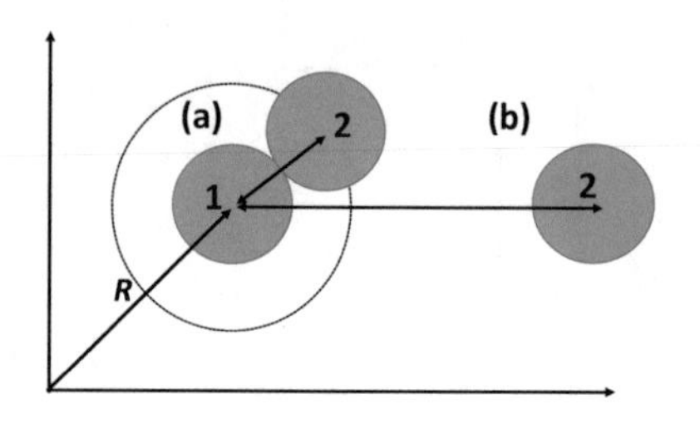

Fig. 2.8 **a** Negative correlation in the region excluded for particle 2, given a particle 1 at *R*. **b** Positive correlation outside the region excluded for particle 2, given a particle 1 at *R*

$$P(R_2|R_1) = 0 \tag{2.59}$$

In terms of "information" we may say that in this region, if we *know* that a particle is at R_1, then we also *know* that no other particle can be at R_2, with $r = |R_2 - R_1|$ smaller than σ, Fig. 2.8.

The second region $\sigma \leq r \leq \sigma + \delta$, is the region of *positive* correlation, i.e.

$$P(R_2|R_1) = g(R_1, R_2)P(R_2) > P(R_2) \tag{2.60}$$

In this case *knowing* that a particle is at R_1, gives us the information that the probability of finding another particle at R_2 (at a distance $\sigma \leq r \leq \sigma + \delta$) is *larger* than the probability $P(R_2)$, Fig. 2.8.

Recall that the MI is always positive, even though the correlations might be either positive or negative. We next turn to the calculation of the entropy-change ΔS (for turning on the interaction) in two steps.

First, we "turn-on" the hard-core interaction (i.e. $\varepsilon = 0$ in Eq. (2.56).In this case:

$$\begin{aligned}
\Delta A &= -\frac{TN(N-1)}{2V^2}\int f_{12}(R_1, R_2)dR_1dR_2 \\
&= -\frac{TN(N-1)}{2V^2}V\int_0^\infty (\exp[-\beta U]-1)4\pi r^2 dr \\
&= \frac{TN(N-1)}{2V}\frac{4\pi\sigma^3}{3}
\end{aligned} \tag{2.61}$$

The corresponding entropy-change is:

$$\Delta S = -\frac{N(N-1)}{2V}\frac{4\pi\sigma^3}{3} < 0 \tag{2.62}$$

This may also be written as:

$$\Delta S \approx N\ln\left[1 - \frac{(N-1)}{V}\frac{4\pi\sigma^3}{6}\right] = N\ln\left[\frac{V - V^{EX}}{V}\right] \tag{2.63}$$

In the last form on the right-hand-side of (2.63) the entropy-change is interpreted in terms of the reduction in the accessible volume for each particle from V to the new volume $V - V^{EX}$, where V^{EX} is the excluded volume for each particle

$$V^{EX} = (N-1)\frac{4\pi\sigma^3}{6} \tag{2.64}$$

The interpretation of the entropy-change in (2.62) is quite clear. When particles have hard-core diameter they produce an *excluded volume* for all other particles. This means that *knowing* the location of one particle, tells us that the probability of finding any other particle in the excluded volume is zero. Thus, whenever we turn on the hard-core repulsive part of the potential, the entropy of the system must decrease, or equivalently the MI is positive.

Next, we turn on the "soft" part of the interactions. In this case the integral in Eq. (2.61) will have two terms:

$$\begin{aligned}\Delta A &= \frac{-TN(N-1)}{2V}\left[-\frac{4\pi\sigma^3}{3} + \int_{\sigma}^{\sigma+\delta}\left[\exp(\beta\varepsilon)-1\right]4\pi r^2 dr\right]\\ &= \frac{TN(N-1)}{2V}\left[\frac{4\pi\sigma^3}{3} - (\exp(\beta\varepsilon)-1)\left(\frac{4\pi(\sigma+\delta)^3}{3} - \frac{4\pi\sigma^3}{3}\right)\right]\end{aligned} \tag{2.65}$$

And the corresponding entropy-change is:

$$\Delta S = -\frac{\partial \Delta A}{\partial T} < 0 \tag{2.66}$$

In this case the entropy-change, ΔS (as well as the corresponding MI), has the two terms: The first, is due to the excluded volume of the particles (negative correlation), and the second, is due to the "soft" interaction. The second term is interpreted as follows: If we know the locations of one or more particles we also know that there is higher probability of finding the other particles in some regions around the first group of particles.

Suppose that we know that one particle is at the location R_1. This information tells us something about the probability of the location of the second particle. If there were no interactions, then:

$$P(R_2) = \frac{1}{V} \tag{2.67}$$

When we "turn-on" the interactions we have:

$$P(R_2|R_1) = \begin{cases} 0 & for r = |R_2 - R_1| \le \sigma \\ > P(R_2) & for \sigma \le r \le \sigma + \delta \end{cases} \tag{2.68}$$

This means that the conditional probability is *smaller* than $P(R_2)$ for $r \leq \sigma$, and it is *larger* than $P(R_2)$ for $\sigma \leq r \leq \sigma + \delta$. In other words, we have both negative and positive correlations, but the MI is always *positive*. We shall study this MI in more details in 1D systems in Sect. 2.7.

2.4 Entropy-Change in Phase Transition

In this section, we shall discuss the entropy-changes associated with phase transitions. Here, by entropy we mean thermodynamic entropy, the units of which are cal/(deg mol). However, as we have seen in Chap. 5 of Ben-Naim [1]. The entropy is up to a multiplicative constant an SMI defined on the distribution of locations and velocities (or momenta) of all particles in the system at equilibrium. To convert from entropy to SMI one has to divide the entropy by the factor $k_B \log_e 2$, where k_B is the Boltzmann constant, and $\log_e 2$ is the natural logarithm of 2, which we denote by ln2. Once we do this conversion from entropy to SMI we obtain the SMI in units of bits. In this section we shall discuss mainly the transitions between gases, liquids and solids. Figure 2.9 shows a typical *phase diagram* of a one-component system. For more details on phase diagrams, see Ben-Naim and Casadei [8].

It is well-known that solid has a lower entropy than liquid, and liquid has a lower entropy of a gas. These facts are usually interpreted in terms of order–disorder. This interpretation of entropy is invalid; more on this in Ben-Naim [6]. Although, it is true that a solid is viewed as more ordered than liquid, it is difficult to argue that a liquid is more ordered or less ordered than a gas.

In the following we shall interpret entropy as an SMI, and different entropies in terms of different MI due to different *intermolecular interactions*. We shall discuss changes of phases at constant temperature. Therefore, all changes in SMI (hence, in entropy) will be due to *locational distributions*; no changes in the momenta distribution.

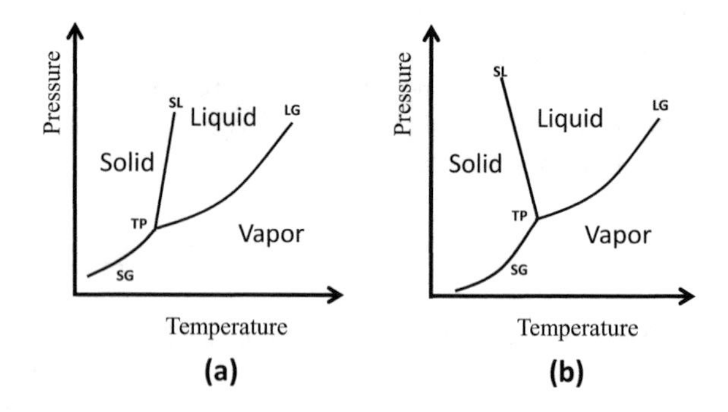

Fig. 2.9 Comparison of the phase diagram of a normal liquid (**a**), and of water, (**b**). Note that the most outstanding difference, is in slopes of the solid–liquid (SL) equilibrium lines; on is positive and the second is negative. The triple point (TP) is the point where the three phases coexist at equilibrium

2.4.1 Solid–Gas Transition

The line SG in Fig. 2.9 is the line along in which solid and gas coexist. The slope of this curve is given by:

$$\left(\frac{dP}{dT}\right)_{eq} = \frac{\Delta S_s}{\Delta V_s} \tag{2.69}$$

In the process of sublimation (s), the entropy-change and the volume change for both are always positive. We denoted by ΔV_s the change in the volume of one mole of the substance, when it is transferred from the solid to the gaseous phase. This volume change is always positive. The reason is that a mole of the substance occupies a much larger volume in the gaseous phase than in the liquid phase (at the same temperature and pressure).

The entropy-change ΔS_s is also positive. This entropy-change is traditionally interpreted in terms of transition from an ordered phase (solid) to a disordered (gaseous) phase. However, the more correct interpretation is that the entropy-change is due to two factors; the huge increase in the *accessible volume* available to each particle and the decrease in the extent of the intermolecular interaction. Note that the slope of the SG curve is quite small (but positive) due to the large ΔV_s.

2.4.2 Liquid–Gas Transition

The line LG in Fig. 2.9 is the line along which liquid and gaseous phases coexist. In this case the slope of the curve is also positive. Both the entropy and the volume change in the vaporization processes are positive. Again, traditionally the entropy of vaporization is interpreted in terms of order–disorder. This interpretation is unfortunately incorrect. It is difficult to argue that gas is more disordered than liquid. The correct interpretation is again the change in the accessible volume per particle, and the weakening of the average intermolecular interactions in the process of the evaporation, hence $\Delta S_v > 0$.

It is worthwhile to mention an empirical law known as the Trouton Law. It states that the *entropy of vaporization* at one atmospheric pressure of many liquids is almost constant;

$$\Delta S_v \approx 85 - 87/Jmol^{-1}K^{-1} \tag{2.70}$$

Table 2.1 shows a few values of the entropy of vaporization. Note that the values of ΔS_v for liquids with strong interactions, such as water, ethanol, and methanol, are much larger than the values for the other liquids.

As one can see from this Table 2.1, the Trouton Law is not really a law. Table 2.1 shows that in many cases the transition from the liquid to the gaseous phase the

Table 2.1 Entropies of vaporization of liquids at their normal boiling point

	$\Delta S_v/\text{Jmol}^{-1}\text{K}^{-1}$
Benzene	87.1
Carbon disulfide	83.7
Carbon tetrachloride	85.8
Cyclohexane	85.1
Dimethyl ether	86.0
Methane	73.2
Methanol	104.1
Ethanol	110.0
Water	109.1
Heavy water	110.8

dominant factor that determines the change in entropy is the change in the accessible volume. However, when there exist very strong intermolecular interactions, the entropy of vaporization ΔS_v becomes very large due to the weakening of the average interactions among the particles in the process of vaporization. As can be seen in Table 2.1 the values of ΔS_v are much larger for liquids with stronger interactions (hydrogen bonds, see Sect. 2.5 below).

2.4.3 Solid–Liquid Transition

The solid–liquid coexistence curve is denoted by SL in Fig. 2.9. In the transition from the solid to the liquid, the entropy-change is always positive. Again, this is traditionally interpreted in terms of transition from order to disorder phases. Although it is certainly true that a solid is more ordered than a liquid, this fact has nothing to do with the positive change in entropy. The change in entropy is due to the change in the total interaction energy among the particles. Unlike the transition from solid to gas where there is a huge change in volume, in the solid to liquid phase the change in volume is usually quite small. This is the reason for the large slope of the curve LG in Fig. 2.9a.

The slope is given by the quotation:

$$\left(\frac{dP}{dT}\right)_{eq} = \frac{\Delta S_m}{\Delta V_m} \tag{2.71}$$

where ΔS_m and ΔV_m are the change in the molar entropy and the volume in the process of melting (m). It should be noted that ΔS_m is always positive, and ΔV_m is positive for most substances. An anomalous case is water for which $\Delta V_m < 0$, i.e. the molar volume of the liquid is smaller than that of ice. In this case the slope of the SL curve is negative, see Fig. 2.9b.

2.5 Liquid Water

Water is known to be a *structured* liquid. However, there is no agreement on how to define the structure of water, see Ben-Naim [9, 10]. Look at Table 2.1, the entropy of vaporization of water is larger than the value expected from Trouton's rule. Also, we see that the entropy of vaporization of heavy water (D_2O) is slightly larger than water (H_2O). This is consistent with the common view that heavy water is a more *structured* liquid than water. However, we can see in Table 2.1 that ethanol has almost the same entropy of vaporization as heavy water though it is difficult to claim that ethanol is more *structured* than either water or heavy water.

Tables of standard entropy are available for many liquids as well as for water and heavy water. It is not easy to compare values of standard entropies of different substances with different degrees of freedom such as vibration, rotation and electronic.

In this chapter we have interpreted the entropy values of a simple liquid in terms of the MI associated with the correlation functions which in turn is associated with the strength of the molecular interactions.

It is usually assumes that H_2O and D_2O have approximately the same internal degrees of freedom. It follows that the higher the entropy of vaporization of D_2O compared with H_2O is due to stronger intermolecular interactions. In this case the main part of the interactions is due to hydrogen bonding, see Ben-Naim [9, 10].

Another measure of the "structure" or the extent of intermolecular interactions in the liquid is the *entropy* of *solvation*. The solvation process is depicted in Fig. 2.10. A single solute molecule (s) is transferred from a fixed position in an ideal gas phase into a fixed position in an ideal gas phase. Figure 2.11 shows some values of the self-solvation entropy of H_2O and D_2O at several temperatures (self-solvation is the process of solvation of a molecule in its own liquid). In all cases we see that ΔS^* of D_2O is more negative than the corresponding value of H_2O.

It is tradition to interpret these values in terms of structural effects (or ordering). Within our interpretation of entropy as a special case of SMI we view the difference in the values of ΔS^* in H_2O and D_2O due to the stronger interaction between D_2O molecules compared with H_2O molecules.

In Appendix, we derive a relationship between the entropy of solvation of a solute s in a solvent in terms of difference in SMI. In the next section we also discuss the solvation entropy of inert gases in water. Here however, we discuss the solvation entropy of H_2O in pure H_2O, i.e. the "solute" is also a water molecule. This is sometimes called self-solvation, i.e. in Fig. 2.10 instead of a solute s inserted in water, we insert a water molecule into pure water. If we do this process at constant temperature T and volume V, the solvation entropy energy is given by:

$$\Delta S_w^* = (k_B \ln 2)[\mathrm{SMI}(N|R_s) - \mathrm{SMI}(N)] \tag{2.72}$$

The factor $(k_B \ln 2)$ is the conversion factor between entropy and SMI. Note that in Eq. (2.72), $\mathrm{SMI}(N)$ is the SMI for N water molecules based on the probability

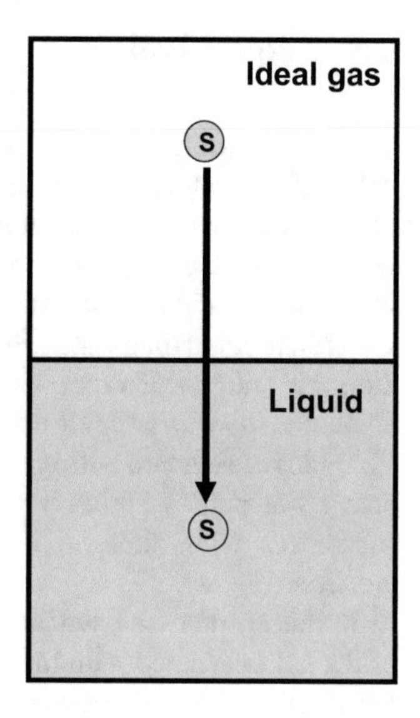

Fig. 2.10 The process of Solvation. A solute molecule (s) is transferred from a fixed position in an ideal gas to a fixed position in the liquid

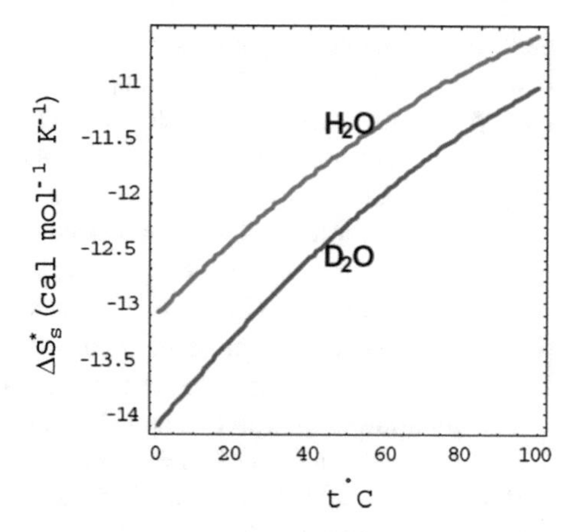

Fig. 2.11 The self-solvation entropy of light and heavy water, as a function of temperature

distribution of locations and orientations of all the N water molecules. Note that since we do the process at constant temperature the momentum SMI does not change in the process. Similarly, SMI($N|R_s$) is the SMI of N solvent molecules and one molecule at a fixed position R_s.

As we can see from Fig. 2.11, the entropy of solvation of water in water is large and negative. This means that when we put a water molecule at a fixed position and orientation in a system of N water molecules, the SMI of the N molecules *decrease.*

The entropy of solvation of many alcohols and other compounds are also large and negative. In Sect. 2.6, we also discuss the solvation entropy of inert solutes in water, for which an interesting explanation for the large negative entropy solvation is available.

As we noted earlier this decrease (in entropy, or in SMI) is traditionally interpreted in terms of the "structure" or "order." From Fig. 2.11 we see that the entropy of solvation of D_2O in D_2O is more negative than the solvation of H_2O in H_2O. This is consistent with the estimates that the intermolecular interactions are stronger in D_2O than in H_2O.

2.6 Aqueous Solutions of Inert Gases

The thermodynamics of aqueous solutions of inert gases involves a few, very exciting and mysterious problems. We shall discuss in this section only one aspect of these systems; the solvation entropy of an inert solute, say, argon in water.

The solubilities of inert solutes such as argon, neon, methane and the like are very small. In the early 1930s and 1940s the data available on the solubility of these solutes in water was very inaccurate. It was known that the solubility of these solutes in water is much smaller than in other organic liquid.

The entropy of solvation (previously referred to as the entropy of solution) of these solutes could be obtained only from very accurate data on the solubility and its dependence on temperature. For more details, see Ben-Naim [11, 12]. It was known that the entropy of solvation of inert solutes in water is large and negative compared with the solvation entropy of the same solutes in typical organic solvents. A few examples are shown in Table 2.2.

In 1945, Frank and Evans [13] published a very influential article on the thermodynamics of solvation of inert solutes in water and in other liquids. They noted that the entropy of solvation of these solutes is much larger and negative in water as compared with the entropy of solvation of the same solutes in other liquids. To explain these findings, the authors *conjectured* that when an inert solute dissolves in water it forms, or builds some kind of structure, which the called "*icebergs,*" around it. This idea was revolutionary at that time. It has captured the imagination of many scientists for more than half a century. How can an inert solute, weakly interacting with water molecules, form an "iceberg"? Frank and Evans did not offer any *proof* that an inert solute builds up iceberg around it, nor did they provide any explanation as to why inert solute should form icebergs. All they did was to interpret the negative change in entropy in terms of increasing the *order*, or equivalently increasing the *structure* of water. Yet, this idea was not only accepted by, but used by many scientists to explain the entropy and the enthalpy of solvation of the non-polar solute in water.

The truth is that Frank and Evans did not contribute anything to understanding the entropy of solvation of inert gases in water. The last statement might be shocking to many chemists who believe that Frank and Evans actually solved the problem. Unfortunately, they did not. Entropy at that time was viewed (and still is) as a measure of the extent of order or disorder in the system. "Structure" is another word for order. Therefore, Frank and Evans suggestion was nothing but the translation of the experiment fact about the negative entropy of solvation into the language of

Table 2.2 Values of the solvation Gibbs energy, entropy, and enthalpy of methane in water and in some non-aqueous solvents at two temperatures

Solvent	t(°C)	ΔG_S^* (cal mol^{-1})	ΔS_S^* (cal mol^{-1} K^{-1})	ΔH_S^* (cal mol^{-1})
Water	10	1747	−18.3	−3430
	25	2000	−15.5	−2610
Heavy water (D_2O)	10	1703	−19.2	−3740
	25	1971	−16.5	−2940
Methanol	10	343	−2.6	−390
	25	390	−3.7	−710
Ethanol	10	330	−3.2	−570
	25	380	−3.5	−650
1-Propanol	10	345	−4.3	−880
	25	400	−3.0	−500
1-Butanol	10	369	−2.8	−420
	25	430	−4.5	−910
1-Pentanol	10	399	−3.3	−530
	25	450	−3.6	−630
1,4-Dioxane	10	538	−0.8	+ 310
	25	553	−1.1	+ 220
Cyclohexane	10	154	−1.9	−390
	25	179	−1.4	−230

order–disorder. Thus, negative ΔS_s^* is equivalent to more order, or more structure, or picturesquely formation of icebergs. For more details, see Ben-Naim [9].

Now, it is known that entropy is *not* a measure of disorder, and even if that were so, just saying that argon creates icebergs (or order, or structure) is not an explanation of the phenomenon. In other words, the question remains as to *why* should inert solutes such as argon create any type of structure when inserted into water. For answers see Ben-Naim [9].

2.7 Entropy and Mutual Information in One Dimensional Fluids

In this section we explore the relationship between the entropy of one-dimensional (1D) fluids and the mutual information due to intermolecular interactions. We start with the general statistical mechanical expression for the entropy of a 1D fluid, and interpret it in terms of Shannon's measure of information (SMI). Next, we present a few numerical example of systems such as hard rods (HR) particles, "square-well"

(SW) interacting particles, and hydrogen-bond-like particles, which exhibit water-like behavior, see Fig. 2.7. It will be concluded that whenever we "turn on" the interactions, the entropy of the system decreases. This may be interpreted in terms of Mutual Information (MI).

It is well known that the partition function, and hence, all the thermodynamic quantities of a one-dimensional (1D) liquid can be written explicitly in terms of the inter-particle interactions, see Ben-Naim [7]. In previous sections we derived a general expression for the entropy of a system of interacting particles. We have seen that when we start with an ideal gas and "turn on" the interactions among all the particles the entropy of the system will always decrease. We interpreted these results in terms of the mutual information (MI) associated with the interactions among the particles. In this chapter we explore the effect of "turning-on" the interactions in a 1D system on the Shannon measure of information (SMI). The 1D system has one important advantage; the SMI (as well as the entropy) may be expressed in terms of a simple one parameter function.

First, we derive a new expression for the entropy of a 1D fluid in terms of the nearest-neighbor distribution function. We then explore the "informational content" of this distribution as well as the corresponding mutual information for fluids of interacting particles via hard-sphere, square-well, and hydrogen-bond like pair interactions, Fig. 2.7.

2.7.1 The General Expression for the Entropy of a 1D Fluid

In Sect. 2.2 we derived a general expression for the change in the entropy of a system when the intermolecular interactions are turned on:

$$\Delta S = -\int Pr\left(R^N\right) \log g\left(R^N\right) dR^N = -I(1; 2; \ldots; N) \tag{2.73}$$

Here, $Pr\left(R^N\right)$ is the probability density of finding the system of N simple-spherical particles at a *specific* configuration $R^N = R_1, R_2, \cdots, R_N$, where R_i is the location of particle i. The correlation function among N particles defined by:

$$g\left(R^N\right) = \frac{Pr\left(R^N\right)}{\prod_{i=1}^{N} Pr(R_i)} \tag{2.74}$$

where $Pr(R_i)dR_i$ is the probability of finding particle i within the element of volume $dR_i = (dx_i dy_i dz_i)$ at R_i.

If we take the Boltzmann constant k_B equal to 1, and use the logarithm to the base 2, we may identify ΔS with the mutual information (MI) among all the particles. Since the MI must always be positive the change in entropy in Eq. (2.73) is always negative.

For a 1D system of particles the expression for the entropy of the system is much simpler. We skip the derivation here and present the final result (assuming that $k_B = 1$ and the logarithm is to the base 2). For details, see Ben-Naim [7, 12].

$$S = -N \log \Lambda - NT \frac{\partial \log \Lambda}{\partial T} - N \int_0^\infty Pr(r) \log Pr(r) dr \tag{2.75}$$

In Eq. (2.75), Λ is the 1D momentum partition function (PF) of the system, defined by:

$$\Lambda = \frac{h}{\sqrt{2\pi m k_B T}} \tag{2.76}$$

where h is the Planck constant, k_B the Boltzmann constant, m the mass of the particles and T the absolute temperature.

$Pr(r)dr$ is the probability of finding a pair of nearest-neighbor particles at a distance between r and $r + dr$. This distribution density is given by:

$$Pr(r) = \frac{\exp[-\beta pr - \beta U(r)]}{\int \exp[-\beta pr - \beta U(r)] dr} \tag{2.77}$$

where, p is the pressure, $\beta = \frac{1}{T}$ (note that $k_B = 1$), and $U(r)$ is the pair interaction potential.

Since the momentum PF depends only on the temperature, the first two terms on the right hand side of Eq. (2.75) will cancel out when we form the difference in entropy:

$$\Delta S = S^l - S^{ig} = -N \int_0^\infty Pr(r) \log Pr(r) dr + N \int_0^\infty Pr^{ig}(r) \log Pr^{ig}(r) dr \tag{2.78}$$

where $Pr^{ig}(r)$ is the same as $Pr(r)$, at the same temperature and pressure, but when all the interactions are "turned off." S^l and S^{ig} are the entropy of the system (with interaction), and the ideal gas (without the interaction), respectively. As we can see from Eq. (2.78) the difference in the entropy of the system when the interactions are on and off, is essentially a difference between two SMI for a system with or without the interactions. According to Eq. (2.73) this difference may be interpreted as a MI among the particles (provided we use $k_B = 1$, and the logarithm to the base 2). It should be noted that in Eq. (2.78) we expressed the entropy-change as a function of one-parameter distribution function $Pr(r)$. This is a huge simplification from Eq. (2.73) in which a multi-dimensional function features.

2.7.2 The General Behavior of the Probability Density $Pr(r)$

The distribution of momenta determines the contribution of the first two terms on the right hand side of Eq. (2.73) to the entropy of the system. The second distribution $Pr(r)$ determines the entropy associated with the locations and the interactions among the particles. In this section we explore the dependence of the probability density $Pr(r)$ on the various parameters of the system.

For ideal gas, $U(r) = 0$, and the distribution density reduces to:

$$Pr(r) = \frac{\exp[-\beta pr]}{\int_0^\infty \exp[-\beta pr]dr} = \frac{\exp[-\beta pr]}{(\beta p)^{-1}} \tag{2.79}$$

Since for an ideal gas the equation of state is:

$$\beta p = \rho = \frac{N}{V} \tag{2.80}$$

We can rewrite the probability density as:

$$Pr(r) = \rho \exp[-\rho r] \tag{2.81}$$

Figure 2.12 shows the probability density Pr for ideal gas at different densities and at different temperatures T, and constant pressure, P. In all cases we have an exponential distribution. As expected the distribution becomes sharper as the density increases or as the temperature decreases, at constant pressure. Note that sharper distribution means smaller SMI.

In Fig. 2.13 we show the distribution density for hard rods with diameter $\sigma = 1$ and pair potential:

$$U_{HS}(r) = \begin{cases} \infty, & \text{for } r < \sigma \\ 0, & \text{for } r \geq \sigma \end{cases} \tag{2.82}$$

The general form of the curves is similar to that of ideal gas. In both cases we note that the distribution becomes more uniform as the temperature increases, or the pressure decreases (or equivalently when the density increases). We will see in the next section that this trend makes the entropy, or the SMI of the system, larger as the density becomes more uniform.

Next, we examine the case of square-well (SW) interaction, defined by:

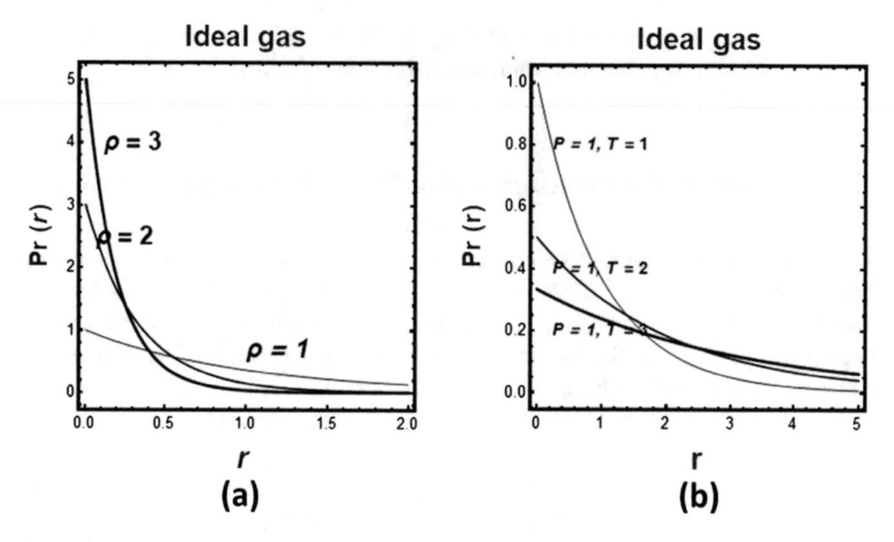

Fig. 2.12 The distribution of nearest-neighbors distance defined in Eq. (2.81) for a 1D ideal gas; **a** at different densities, **b** at different temperatures and constant pressure

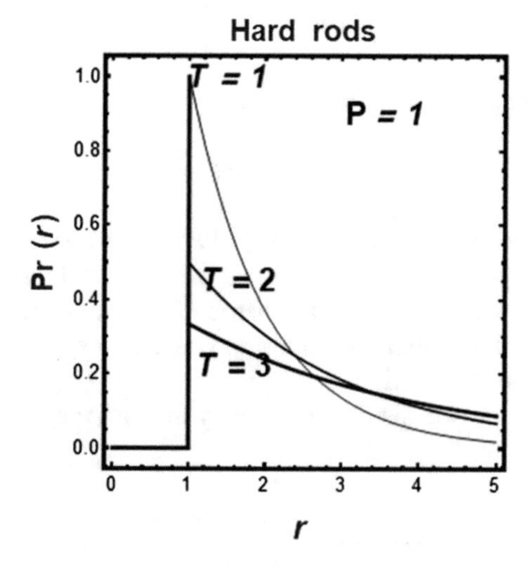

Fig. 2.13 The distribution of nearest-neighbors distance defined in Eq. (2.81) for the hard-rod pair-potential at three different temperatures and constant pressure

$$U_{sw}(r) = \begin{cases} \infty, & \text{for } r < \sigma \\ -\varepsilon, & \text{for } \sigma < r < \sigma < \delta \\ 0, & \text{for } r > \sigma + \delta \end{cases} \tag{2.83}$$

Here, ε measures the strength, and δ the range of the interaction energy.

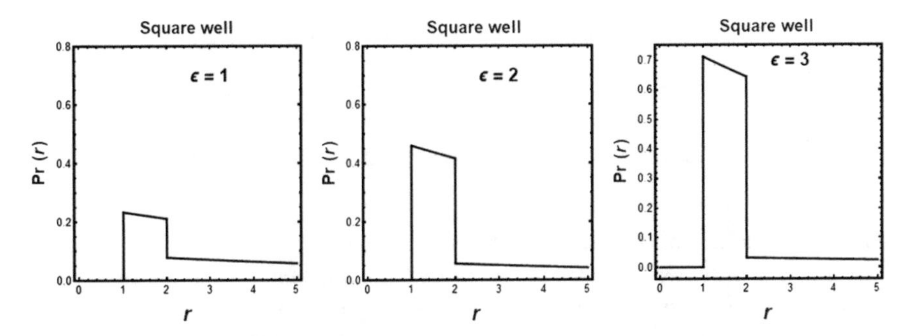

Fig. 2.14 The distribution of nearest-neighbors distance, defined in Eq. (2.81) for the square-well pair-potential for different values of ε

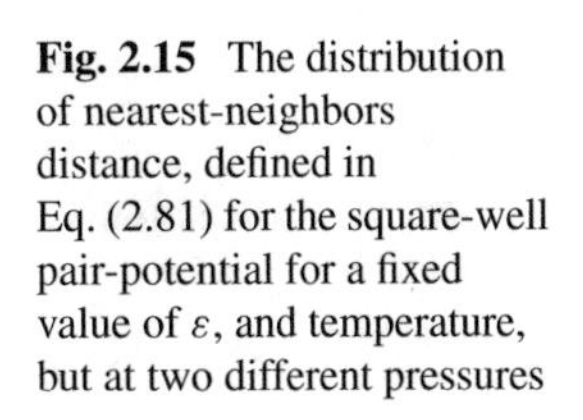
Fig. 2.15 The distribution of nearest-neighbors distance, defined in Eq. (2.81) for the square-well pair-potential for a fixed value of ε, and temperature, but at two different pressures

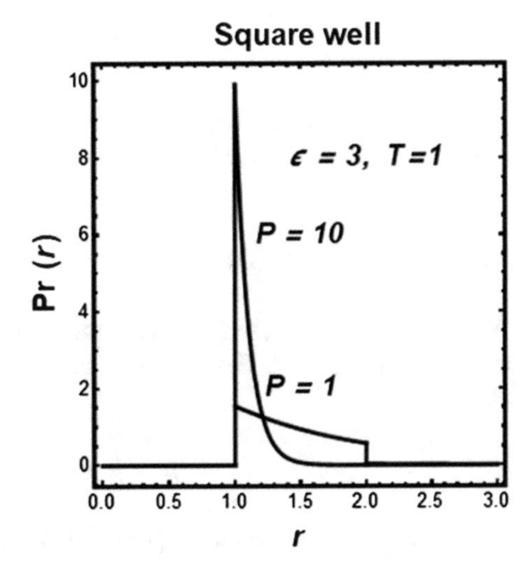

Figure 2.14 shows the distribution density for this pair potential with different values of ε. Figure 2.15 shows the same distribution but at two different pressures. As expected when the pressure become very high we get a sharp distribution at the distance, $r = 1$, equal to the length of the particles.

2.7.3 The Entropy-Change Due to Turning-On the Interaction Energy

In this section we study a few examples of the effect of "turning-on" the interaction energy on the entropy of the system. As we already know from the previous sections turning on the interaction energy, at constant *T, P* always reduces the entropy of the system, or equivalently, increases in the SMI. Before we show a few examples it is

(a)

(b)

Fig. 2.16 **a** A 1-D system of hard rods, and **b** close-packed system of hard rods

important to note that one should be careful in choosing the parameters p, T, and ε of the system (here, $\sigma = \delta = 1$ in all of the examples). When the pressure is too high or the temperature too low, the system might reach a close-packed density, Fig. 2.16, beyond which we might get unrealistic values for the entropy-changes, as well as for other thermodynamic quantities. For instance, for hard-rod particles the exact equation of state is:

$$\beta p = \frac{\rho}{1 - \rho\sigma} \tag{2.84}$$

Thus, the pressure diverges when $\rho\sigma = 1$, for $\rho\sigma > 1$ we obtain a negative pressure, which is physically meaningless.

2.7.3.1 Hard Rods (HR)

We start with the a system of hard rod particles with interaction energy potential as shown in Fig. 2.7a. For this case we have an exact expression for the chemical potential:

$$\mu = T\ln(\Lambda p/T) + \sigma p \tag{2.85}$$

The change in the chemical potential is simply:

$$\Delta\mu = \mu - \mu^{ig} = \sigma p \tag{2.86}$$

Thus, for the HR of diameter $\sigma = 1$, $\Delta\mu = p$, the corresponding entropy-change is:

$$\Delta S = S - S^{ig} = 0 \tag{2.87}$$

This result is valid for the process of "turning on" the interaction at p and T constants. However, when the "turning on" of the interactions are carried out at constant volume, or constant density, the change in the entropy is always negative. The exact expression for the change in Gibbs energy per mole of particles at constant T and ρ is:

$$\Delta\mu = -T\ln(1 - \rho\sigma) + \sigma T\frac{\rho}{1 - \rho\sigma} \tag{2.88}$$

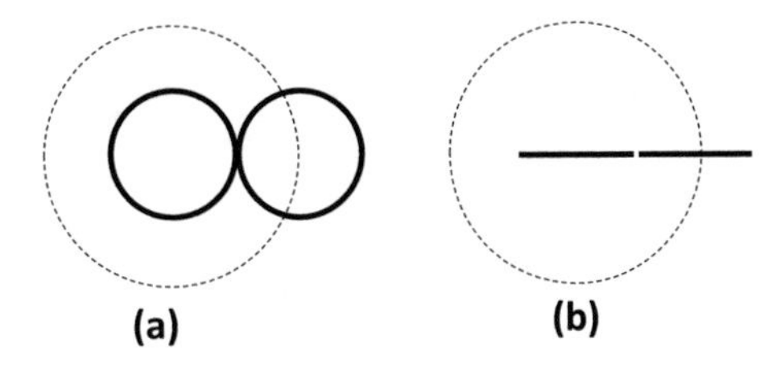

Fig. 2.17 **a** The excluded volume by a hard sphere, and **b** by a hard rod

Clearly, this expression diverges for $\rho\sigma \geq 1$ (when the density is larger than the close-packed density). For very low densities we have:

$$\Delta\mu \approx T\rho\sigma + T\rho\sigma = 2T\rho\sigma \tag{2.89}$$

Hence, the corresponding entropy-change is:

$$\Delta S = -2\rho\sigma \tag{2.90}$$

which is always *negative*. The reason for this particular value is that the "volume" accessible to each particle is reduced from L to L -2σ, where 2σ is the excluded "volume" per particle, Fig. 2.17.

Note that for very small density when $\rho\sigma \ll 1$ we can rewrite expression (2.88) approximately as:

$$\Delta S \approx \ln(1 - 2\rho\sigma) = \ln\left(\frac{V - 2N\sigma}{V}\right) \tag{2.91}$$

which is the change in entropy due to the transition from V to $V - 2N\sigma$.

2.7.3.2 Square-Well (SW) Potential

We next explore the behavior of a system interacting by the square well interaction energy as in Fig. 2.7b. For all the following discussions we choose $\sigma = \delta = 1$, and examining the dependence of ΔS on T, p, and the strength of the interaction ε.

Figure 2.18a shows the values of ΔS as a function of T for different values of ε. As ε increases the average interaction energy between pair of nearest-neighbor particles increase. This is turn causes an increase in the MI, hence larger (and negative) values of ΔS. For any value of ε, when the temperature increases the correlation between the locations of the particles *decrease*, hence, also the MI and the ΔS tends to zero.

Similar data is shown in Fig. 2.18b. Here, ΔS is plotted as a function of p for different values of ε. Note that as the pressure tends to zero, the system tends to an ideal gas, hence, correlations tend to unity and the corresponding MI and ΔS tend to zero.

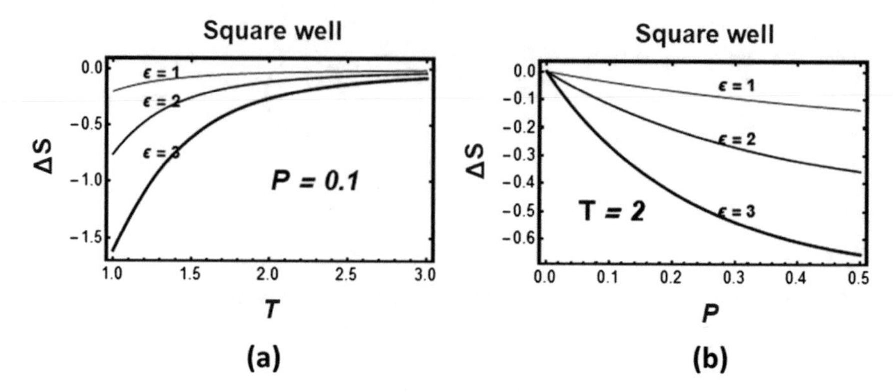

Fig. 2.18 The entropy-change defined for the square-well pair-potential for different values of ε. **a** As a function of T, and **b** as a function of p

It should be noted that the parameters for all these illustrations should be chosen with care. If one chooses too large pressures or too low temperatures the entropy of the system becomes negative, see Fig. 2.19, which is not realistic. The reason for these unrealistic results is that for some parameters we might get into densities higher than the close-packed one.

Finally, we show in Fig. 2.20 the dependence of ΔS on ε, for different values of p, and T. As it is clear for all cases the entropy-change is always negative and becomes larger the larger the strength of the interactions. At very large values of ε, the system reaches a constant value.

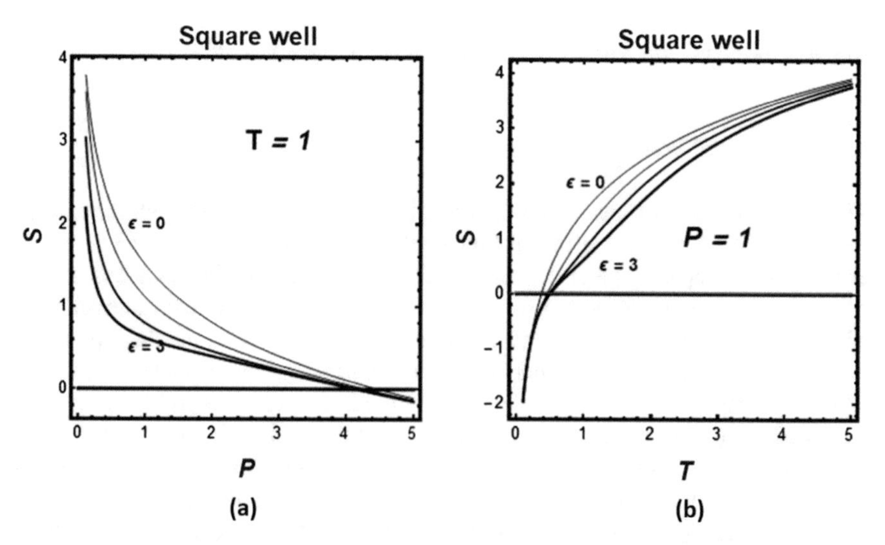

Fig. 2.19 The entropy for the square-well pair-potential for different values of ε. **a** As a function of p, and **b** as a function of T. Note the negative values of the entropy at very high pressures or low temperatures. These values correspond to system for which the pressure becomes negative

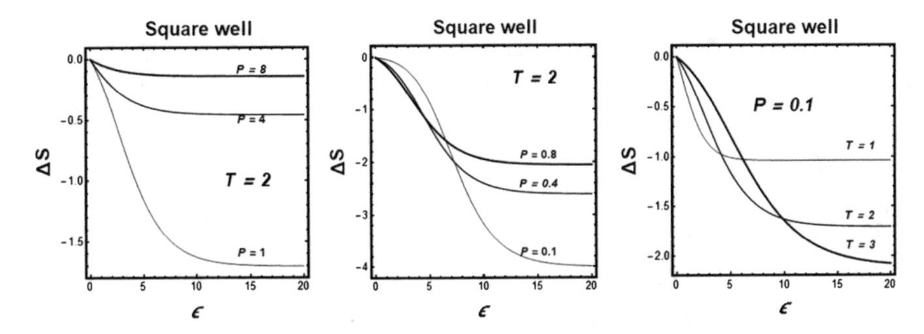

Fig. 2.20 The entropy-change for the square-well pair-potential as a function of ε at different pressures and temperatures

2.7.3.3 Water-Like Particles in 1D

Water-like particles were originally studied in a 1D system in a lattice model and in a liquid, See Ben-Naim [9, 10, 12]. The pair potential is shown in Fig. 2.7c. This potential was designed in such a way as to satisfy the "water-principle" (see Ben-Naim [9, 10]), i.e. the existence of regions of low local density and high local density associated with large and small binding energy, respectively.

In 3D models for water the hydrogen-bond like potential is supposed to create *local structure* in the liquid. This in turn was interpreted in terms of "structure of water" hence, the relatively low entropy of water. In the study of the 1D model of water-like particles, one may get some typical anomalous properties of water such as the negative temperature dependence of the volume in some regions in the phase diagram. Thus, while the 1D model can explain some anomalous properties of water, it cannot show the typical 3D structure of liquid water. Therefore, the entropy-change associated with the "turning on" of the interaction energies is a result of the *correlations* in the locations of the particles and not on the "structure" of the liquid.

Figure 2.21 shows some results for the entropy-change in this system. Note that in this system there are two minima in the pair potential, Fig. 2.7c, therefore as one changes either the temperature or the pressures one get into regions of average distances where either the first or the second minimum will determine the extent of correlations, hence also the entropy of the system is not expected to change monotonically with T or with P.

2.7.4 *Conclusion*

We have seen that whenever we "turn on" the interactions between the particles, the entropy of the system always decreases. This may be interpreted in terms of Mutual Information (MI). The reason is that whenever there are interactions between particles there are also correlations in the locations of the particles, and these correlations

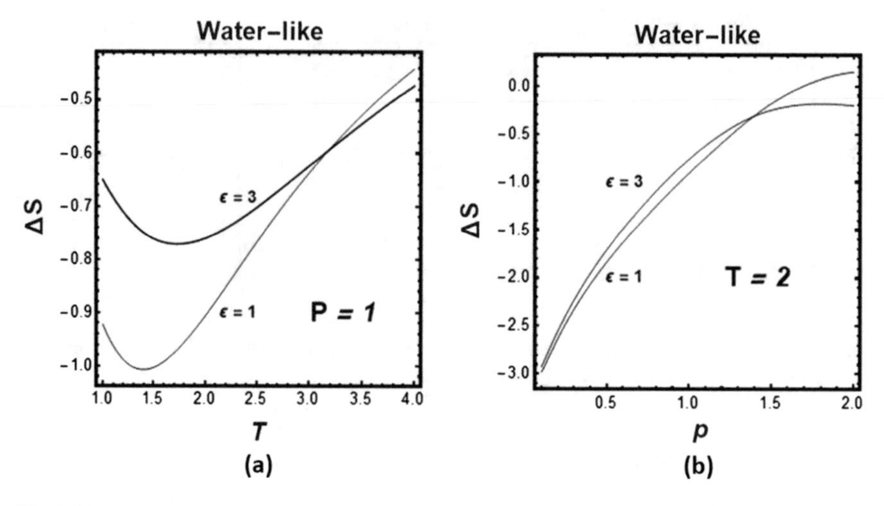

Fig. 2.21 The entropy-change for the water-like pair-potential for different values of ε. **a** As a function of *T*, and **b** as a function of *p*

determine the MI among all the particles in the system. There is no need to invoke either changes in *structure* or changes in *order* to explain entropy-changes.

Appendix 1: Solvation Entropy of a Solute as Difference in SMI

We start with the definition of the solvation Helmholtz energy of a solute s in a solvent containing N solvent molecules at a given temperature T, and volume V.

$$\Delta A_s^* = -k_B T \ln\left\{\int P(\boldsymbol{R}^N)\exp[-\beta B_s]d\boldsymbol{R}^N\right\} \tag{2.92}$$

ΔA_s^* is the change in the Helmholtz energy for the process of transferring a single solute molecule s from a fixed position in the vacuum (or in an ideal gas) into the solvent at a fixed position $\boldsymbol{R}_s$. For simplicity, we assume that both of the solute s and solvent molecules are simple spherical molecules. k_B is the Boltzmann constant. B_s is the total interaction energy of the solute molecule at $\boldsymbol{R}_s$ with all solvent molecules at a configuration $\boldsymbol{R}^N$. $P(\boldsymbol{R}^N)$ is the probability density of finding the configuration $\boldsymbol{R}^N = \boldsymbol{R}_1, \ldots, \boldsymbol{R}_N$ of the solvent molecules, and is given by:

$$P(\boldsymbol{R}^N) = \frac{\exp[-\beta U_N(\boldsymbol{R}^N)]}{\int \exp[-\beta U_N(\boldsymbol{R}^N)]d\boldsymbol{R}^N} \tag{2.93}$$

$U_N(\boldsymbol{R}^N)$ is the total interaction energy among the solvent molecules at the configuration $\boldsymbol{R}^N$. The integration in both integrals (2.92) and (2.93) is over all the configurations of the solvent molecules in the volume V, $\beta = (k_B T)^{-1}$.

Similarly, in a system of N solvent molecules and one solute molecule, the probability density of finding the configuration $\boldsymbol{R}^N$, $\boldsymbol{R}_s$ is given by:

$$P(\boldsymbol{R}^N, \boldsymbol{R}_s) = \frac{\exp[-\beta U_N - \beta B_s]}{\int \exp[-\beta U_N - \beta B_s] d\boldsymbol{R}^N d\boldsymbol{R}_s} \tag{2.94}$$

We shall use the shorthand notations Z_N and Z_{N+1} for the integrals in (2.93) and (2.94), respectively.

The entropy of the solvation is obtained by taking the derivative of ΔA_s^* with respect to temperature:

$$\Delta S_s^* = -\left(\frac{\partial \Delta A_s^*}{\partial T}\right)_{V,N} \tag{2.95}$$

Before we take the derivative in (2.95) we note that Z_{N+1} in (2.94) may be rewritten as:

$$Z_{N+1} = \int \exp[-\beta U_N - \beta B_s] d\boldsymbol{R}^N d\boldsymbol{R}_s = V \int \exp[-\beta U_N - \beta B_s] d\boldsymbol{R}^N \tag{2.96}$$

In (2.96) we defined all the locations $\boldsymbol{R}_i$ of the solvent molecules relative to the location of the solute $\boldsymbol{R}_s$. Hence, we can integrate over $\boldsymbol{R}_s$ to obtain the volume of the system, i.e.

$$V = \int_V d\boldsymbol{R}_s \tag{2.97}$$

We now take the derivative in (2.95) to obtain:

$$\begin{aligned} \Delta S_s^* &= k_B \ln \int P(\boldsymbol{R}^N) \exp[-\beta B_s] d\boldsymbol{R}^N \\ &+ k_B \frac{\int \exp[-\beta U_N - \beta B_s](+\beta(U_N - B_s)) d\boldsymbol{R}^N}{Z_{N+1}/V} \\ &- k_B \frac{\int \exp[-\beta U_N](+\beta U_N) d\boldsymbol{R}^N}{Z_N} \end{aligned} \tag{2.98}$$

We now use the conditional distribution of finding the configuration $\boldsymbol{R}^N$ given a solute molecule at $\boldsymbol{R}_s$.

$$P(\boldsymbol{R}^N | \boldsymbol{R}_s) = \frac{P(\boldsymbol{R}^N, \boldsymbol{R}_s)}{P(\boldsymbol{R}_s)} = V P(\boldsymbol{R}^N, \boldsymbol{R}_s) = V \frac{\exp[-\beta U_N - \beta B_s]}{Z_{N+1}} \tag{2.99}$$

With this conditional probability we can rewrite Eq. (2.98) as:

$$\Delta S_s^* = (k_B \ln 2)\left[-\int P\left(\boldsymbol{R}^N \middle| \boldsymbol{R}_s\right)\log_2 P\left(\boldsymbol{R}^N \middle| \boldsymbol{R}_s\right)d\boldsymbol{R}^N + \int P\left(\boldsymbol{R}^N \middle| \boldsymbol{R}_s\right)\log_2 P\left(\boldsymbol{R}^N\right)d\boldsymbol{R}^N\right]$$
$$= (k_B \ln 2)[\mathrm{SMI}(N|\boldsymbol{R}_s) - \mathrm{SMI}(N)] \quad (2.100)$$

Thus, aside from the constant $(k_B \ln 2)$, the solvation entropy of the solute is simply the difference in the SMI of the N solute particles before, and after putting the solute at $\boldsymbol{R}_s$. Note again that this SMI is based on the distribution of the locations (and in more general cases on the orientations) of all the solvent molecules, the velocity distribution, and the corresponding SMI does not change in this process.

References

1. Ben-Naim, A. (2017). *Information theory, Part I: An introduction to the fundamental concept*. World Scientific.
2. Sackur, O. (1911). *Annalen der Physik, 36*, 958.
3. Tetrode, H. (1912). *Annalen der Physik, 38*, 434.
4. Ben-Naim, A. (2018). An informational theoretical approach to the entropy of liquids and solutions. *Entropy, 20*, 514.
5. Shannon, C. E. (1948). A mathematical theory of communication. *Bell System Technology Journal, 27*, 379.
6. Ben-Naim, A. (2020). *Entropy: The greatest blunder ever in the history of science*. Independent Publisher, Amazon.
7. Ben-Naim, A. (1992). *Statistical thermodynamics for chemists and biochemists*. Plenum Press.
8. Ben-Naim, A., & Casadei, D. (2016). *Modern thermodynamics*. World Scientific.
9. Ben-Naim, A. (2009). *Molecular theory of water and aqueous solutions. Part I: Understanding water*. World Scientific.
10. Ben-Naim, A. (2011). *Molecular theory of water and aqueous solutions. Part II: The role of water in protein folding, self-assembly and molecular recognition*. World Scientific.
11. Ben-Naim, A. (2016). *Myths and verities in protein folding theories*. World Scientific.
12. Ben-Naim, A. (2014). *Statistical thermodynamics, with applications to life sciences*. World Scientific.
13. Frank, H. S., & Evans, M. W. (1945). *The Journal of Chemical Physics, 13*, 507.

Chapter 3
Application of Multivariate Mutual Information to Study Spin Systems

In Chap. 4 of Ben-Naim [1] we defined two different multivariate mutual information (MI). We also studied the case of three spin systems where we introduced the concept of *frustration* in terms of MI. In this chapter we generalized the same study for different numbers of spins.

We shall start with a review of the two different multivariate MI; one is based on the *total correlations* (denoted TI), and the second based on *conditional* information (denoted CI). The reader who is familiar with these terms can skip Sects. 3.1 to 3.5. It should be noted that there are many possibilities of generalizing the MI for large number of random variables (random variables), we shall discuss only two possible generalizations. In this chapter we shall study the MI of spins systems with different numbers of spins and different geometrical arrangements.

3.1 Definition of Multivariate MI Based on Total Correlations

There are several ways of defining the MI for many random variables. The simplest and the most "natural" definition, is the one based on the interpretation of the MI as a measure of the *average correlation* between two random variables.

The MI for two experiments or random variables is defined by:

$$
\begin{aligned}
I(X_1; X_2) &= \sum_{x_1, x_2} p(x_1, x_2) \log \frac{p(x_1, x_2)}{p(x_1)p(x_2)} \\
&= \sum_{x_1, x_2} p(x_1, x_2) \log[g(x_1, x_2)] \qquad (3.1)
\end{aligned}
$$

In Eq. (3.1) we denoted by $p(x_1, x_2)$ the pair-probability, and $g(x_1, x_2)$ is the pair-correlation between the event $(X_1 = x_1)$ and the event $(X_2 = x_2)$. When the

A. Ben-Naim, *Information Theory and Selected Applications*,
https://doi.org/10.1007/978-3-031-21276-5_3

two events are independent, $g(x_1, x_2) = 1$. We say that the two events are *positively correlated* when $g(x_1, x_2) > 1$ (i.e., $\log[g(x_1, x_2)] > 0$). We say that the two events are *negatively correlated* when $g(x_1, x_2) < 1$ (i.e., $\log[g(x_1, x_2)] < 0$). From Eq. (3.1), it follows that $I(X_1; X_2)$ is a measure of the average of the logarithm of the correlation between the two *random variables.*

A generalization of the quantity in Eq. (3.1) for any number of random variables, based on this interpretation of the MI, is straightforward. We first define the correlation function between the n events $(X_1 = x_1), \ldots, (X_n = x_n)$ by:

$$g(x_1, x_2, \ldots, x_n) = \frac{p(x_1, x_2, \ldots, x_n)}{\prod_{i=1}^{n} p(x_i)} \tag{3.2}$$

where, $p(x_1, \ldots, x_n)$ is the probability of finding the event $(X_1 = x_1, \ldots, X_n = x_n)$. Next, we define the *total mutual information* by:

$$TI(X_1; \ldots; X_n) = \sum_{x_1, \ldots, x_n} p(x_1, \ldots, x_n) \log[g(x_1, \ldots, x_n)] \tag{3.3}$$

Clearly, the quantity *TI* is an average of the logarithm of the correlation function between the n events $(X_1 = x_1), \ldots, (X_n = x_n)$. The quantity $g(x_1, \ldots, x_n)$ is sometimes referred to as *total* correlation between the n events. When the n variables are independent, the *total correlation* function is unity, i.e. $p(x_1, \ldots, x_n) = \prod_{i=1}^{n} p(x_i)$, hence $g(x_1, \ldots, x_n) = 1$ [4]. Clearly, the *TI* is always a non-negative quantity. This fact follows from the Kullback–Leibler inequality, see Ben-Naim [1]. It is zero, if and only if, all the n random variables are independent.

3.2 Definition of Multivariate MI Based on *Conditional Information*

Another generalization of the MI due to Fano [2], starts from the identity:

$$I(X_1; X_2) = H(X_1) - H(X_1|X_2) \tag{3.4}$$

As shown in Ben-Naim [1], this quantity is *always positive.* This means that having any information on X_2 can only *reduce* the amount of uncertainty about X_1. Note that when we say "information on X_2 will reduce the uncertainty about X_1," we mean that we *know* the *specific* result of the experiment X_2; this *cannot increase* the uncertainty about X_1. Here, the uncertainty is about the entire set of the outcomes of X_1. Next, we use the two well-known identities, see Ben-Naim [1]:

$$I(X_1; X_1) = H(X_1) \tag{3.5}$$

$$I(X_1; X_1|X_2) = H(X_1|X_2) \tag{3.6}$$

The first equality Eq. (3.5) means that the mutual information between X_1 and X_1 is the same as the SMI of X_1. Clearly, the same is true for the conditional mutual information, Eq. (3.6); the mutual information between X_1 and X_1, given X_2 is the same as the conditional SMI of X_1, given X_2.

We now use the two identities (3.5) and (3.6), to generalize (3.4) for three random variables. We first, we rewrite the MI in (3.4) in terms of the two self-information as defined in (3.5) and (3.6), i.e.:

$$I(X_1; X_2) = I(X_1; X_1) - I(X_1; X_1|X_2) \tag{3.7}$$

For three random variables we generalize Eq. (3.7) and define a new multivariate MI:

$$CI(X_1; X_2; X_3) = I(X_1; X_2) - I(X_1; X_2|X_3) \tag{3.8}$$

We shall refer to the quantity, denoted CI, and defined in Eq. (3.8), as the *conditional* MI between the three random variables. It follows from Eq. (3.8) that $CI(X_1; X_2; X_3)$ is a measure of the effect of knowing X_3 on the MI between X_1 and X_2. Note that the definition in Eq. (3.8) is symmetric with respect to interchanging X_1 and X_2, but not with respect to X_3. However we can show that the conditional MI, is in fact, symmetric with respect to the three variables. This means that $CI(X_1; X_2; X_3)$ is invariant to any permutations of the random variable.

To show this, we start by using the definitions of the MI on the right hand side of Eq. (3.8), which we rewrite as:

$$\begin{aligned} CI(X_1; X_2; X_3) = &\sum_{x_1,x_2} p(x_1, x_2) \log\left[\frac{p(x_1, x_2)}{p(x_1)p(x_2)}\right] \\ &- \sum_{x_3} p(x_3) \sum_{x_1,x_2} p(x_1; x_2|x_3) \log\left[\frac{p(x_1, x_2|x_3)}{p(x_1|x_3)p(x_2|x_3)}\right] \end{aligned} \tag{3.9}$$

The first term on the right-hand-side of Eq. (3.9) is, by definition the MI, $I(X_1; X_2)$. The second term is the average over all values x_3 of the conditional MI between X_1 and X_2 given the *specific* value of x_3.

We recall the definitions of the conditional probabilities:

$$p(x_1|x_3) = \frac{p(x_1, x_3)}{p(x_3)} \tag{3.10}$$

$$p(x_2|x_3) = \frac{p(x_2, x_3)}{p(x_3)} \tag{3.11}$$

$$p(x_1, x_2|x_3) = \frac{p(x_1, x_2, x_3)}{p(x_3)} \tag{3.12}$$

We now use the identity $\sum p(x_3) = 1$, in the first term on the right-hand-side of Eq. (3.9), to rewrite Eq. (3.9) as:

$$CI(X_1; X_2; X_3) = -\sum p(x_1, x_2, x_3)\log\left[\frac{p(x_1, x_2, x_3)p(x_1)p(x_2)p(x_3)}{p(x_1, x_2)p(x_1, x_3)p(x_2, x_3)}\right] \quad (3.13)$$

In this form, we see that $CI(X_1; X_2; X_3)$ is symmetric with respect to all the three variables X_1, X_2 and X_3. We can also rewrite (3.13) as:

$$\begin{aligned} CI(X_1; X_2; X_3) &= I(X_1; X_2) - I(X_1; X_2|X_3) \\ &= I(X_1; X_3) - I(X_1; X_3|X_2) \\ &= I(X_2; X_3) - I(X_2; X_3|X_1) \end{aligned} \quad (3.14)$$

Clearly, the quantity $CI(X_1; X_2; X_3)$ measures the difference between the MI between any two variables, and the conditional MI between the same two variables given the knowledge about the third variable.

3.3 Relationship Between the Conditional MI and the Various SMI

In this section we express $CI(X_1; X_2; X_3)$ in terms of SMIs. We use Eq. (3.13) to rewrite the *CI* as:

$$\begin{aligned} CI(X_1; X_2; X_3) = &-\sum p(x_1, x_2, x_3) \log p(x_1, x_2, x_3) \\ &-\sum p(x_1) \log p(x_1) - \sum p(x_2) \log p(x_2) \\ &-\sum p(x_3) \log p(x_3) \\ &+\sum p(x_1, x_2) \log p(x_1, x_2) + \sum p(x_1, x_3) \log p(x_1, x_3) \\ &+\sum p(x_2, x_3) \log p(x_2, x_3) = H(X_1) + H(X_2) + H(X_3) \\ &- H(X_1, X_2) - H(X_1, X_3) - H(X_2, X_3) + H(X_1, X_2, X_3) \end{aligned} \quad (3.15)$$

In the final form on the right hand side of (3.15), we expressed the quantity $CI(X_1; X_2; X_3)$ in terms of combination of SMIs. This result is reminiscent of the inclusion–exclusion principle in probability theory, except that we have here SMIs instead of probabilities. This form of $CI(X_1; X_2; X_3)$ has motivated Matsuda [3] and others to *define "Higher-order mutual information"* by Eq. (3.15). Also, the commonly used Venn diagram for SMI is motivated by this result. As we noted in Chap. 1, the Venn diagrams are useful in studying probabilities of overlapping events. It is not recommended to use it for the SMI and the MI which measure the extent of *dependence* between events. We shall continue to use the notation *CI* for the quantity defined in Eq. 3.15, although this form does not involve any conditional probabilities.

In probability theory, whenever each of the pairs of events A, B and C are *mutually exclusive*, i.e. disjoint in pairs, $A \cap B = A \cap C = B \cap C = 0$, it follows that the three events *A*, *B* and *C* are also disjoint, i.e. $A \cap B \cap C = 0$. However, the inverse of this statement is not true. Three events may have no point in common, but pairs of the three events are not necessarily disjoint. when the events are *independent* in pairs, it does not follow that they are independent in triplets. Note also that independence in triplets does not imply independence in pairs [see example in Ben-Naim, [4]]. For all these reasons, it is not advisable to use the Venn diagrams for the SMIs in information theory.

As we have seen in Chap. 1 and we shall see in the following sections, there are cases when the three random variables are dependent, yet the triplet MI as defined in (3.9) is *negative*. Presenting such a case by a Venn diagram would require regions of overlapping between SMIs having *negative* area. In studying Venn diagrams, all overlapping events are shown as (positive) areas. If one uses Venn diagrams to describe SMI and MI, one might encounter "negative" areas, which is obviously meaningless. Perhaps, this is the reason why one calls this case *frustration*, see below.

3.4 The Formal Connection Between the TI and CI

We now derive the general relationship between the *TI* and the *CI* for the case of three *random variables*. We start with Eq. (3.15), which we rewrite as:

$$\begin{aligned} CI(X_1; X_2; X_3) &- \sum p(x_1) \log p(x_1) - \sum p(x_2) \log p(x_2) \\ &- \sum p(x_3) \log p(x_3) + \sum p(x_1, x_2) \log p(x_1, x_2) \\ &+ \sum p(x_1, x_3) \log p(x_1, x_3) + \sum p(x_2, x_3) \log p(x_2, x_3) \\ &- \sum p(x_1, x_2, x_3) \log p(x_1, x_2, x_3) \end{aligned} \tag{3.16}$$

Take note also that each singlet probability is the marginal probability of the pair-probability, e.g., $p(x_1) = \sum_{x_2} p(x_1, x_2) = \sum_{x_2, x_3} p(x_1, x_2, x_3)$. Therefore, we can rewrite Eq. (3.16) as:

$$CI(X_1; X_2; X_3) = TI(X_1; X_2) + TI(X_1; X_3) + TI(X_2; X_3) - TI(X_1; X_2; X_3) \tag{3.17}$$

Note that *TI* for two *random variables* is the same as the mutual information $I(X_1; X_2)$.

Thus, we have related the *CI* to the *TI* defined with respect to the total correlation functions. Note also that in Eq. (3.17) $CI(X_1, X_2, X_3)$ is symmetric with respect to the three random variables.

3.5 Reinterpretation of the CI in Terms of MIs

We start with the *CI* as defined in Eq. (3.13), which we can rewrite as:

$$\begin{aligned} CI(X_1; X_2; X_3) &= -\sum p(x_1, x_2, x_3) \\ &\quad \log\left[\frac{p(x_1, x_2, x_3)}{p(x_1)p(x_2, x_3)} \frac{p(x_1)p(x_2)}{p(x_1, x_2)} \frac{p(x_1)p(x_3)}{p(x_1, x_3)}\right] \\ &= \sum p(x_1, x_2, x_3)\log\left[\frac{p(x_1, x_2)}{p(x_1)p(x_2)}\right] \\ &\quad + \sum p(x_1, x_2, x_3)\log\left[\frac{p(x_1, x_3)}{p(x_1)p(x_3)}\right] \\ &\quad - \sum p(x_1, x_2, x_3)\log\left[\frac{p(x_1, x_2, x_3)}{p(x_1)p(x_2, x_3)}\right] \end{aligned} \tag{3.18}$$

Using the definition of the MI between the two random variables we rewrite (3.18) as:

$$CI(X_1; X_2; X_3) = I(X_1; X_2) + I(X_1; X_3) - I(X_1; (X_2, X_3)) \tag{3.19}$$

where the last term, on the right hand side of (3.19) is defined by:

$$I(X_1; (X_2, X_3)) = \sum p(x_1, x_2, x_3)\log\left[\frac{p(x_1, x_2, x_3)}{p(x_1)p(x_2, x_3)}\right] \tag{3.20}$$

In Eq. (3.19) the quantity *CI* is written as the *difference* between the sum of the MIs $I(X_1; X_2) + I(X_1; X_3)$ and the MI between X_1 and the *joint* rv (X_2, X_3).

Note that on the right hand side of (3.19), although we have three random variables, the *mutual information* are only between *two random variables*, i.e. $I(X_1; (X_2, X_3))$ is the amount of information that we gain about X_1 from the knowledge of both X_2 and X_3. This interpretation is useful in applications when one transmits information from two (or more) sources. Qualitatively, we expect that the MI between X_1 and X_2, and the MI between X_1 and X_3, will be *larger* than the MI between X_1 and the joint *rv* (X_2, X_3). However, there are cases when $CI(X_1; X_2; X_3)$ is negative. Perhaps, this is the reason for referring to this case as "frustration" [3]. We shall further discuss this interpretation of CI, as a measure of frustration, in the following sections.

3.6 Generalization to Any N Random Variables

As we have seen in Sect. 3.1, the generalization of the *TI* for any number of random variables is straightforward. The generalization of the *CI* for any number of random variables starts from the definition in (3.14):

$$CI(X_1; X_2; X_3) = I(X_1; X_2) - I(X_1; X_2|X_3) \tag{3.21}$$

For four variables, we write the *CI* as:

$$CI(X_1; X_2; X_3; X_4) = CI(X_1; X_2; X_3) - CI(X_1; X_2; X_3|X_4) \tag{3.22}$$

And for any n as:

$$CI(X_1; X_2; \ldots; X_n) = CI(X_1; X_2; \ldots; X_{n-1}) - CI(X_1; X_2; \ldots; X_{n-1}|X_n) \tag{3.23}$$

This expression can be expanded in terms of the SMIs as follows:

$$CI(X_1; \ldots; X_n) = \sum_{k=1}^{n}(-1)^{k-1} \sum_{(i_1,\ldots,i_k)} H\left(X_{i_1}, \ldots, X_{i_k}\right) \tag{3.24}$$

Note that the second sum on the right hand side of (3.24) is over all possible sets of indices $(i_1, i_2, \ldots, i_k)$, with : $1 < i_1 < i_2 < \ldots i_k < k$.

Again, we note that this expression is reminiscent of the inclusion–exclusion principle in probability theory. This is the reason some authors use the Venn diagram for the conditional MI. Matsuda [3] uses Eq. (3.24) to define the generalized the MI, we prefer Eq. (3.23) which is also the reason we refer to this quantity as the Conditional MI.

3.7 Some Properties of the Multivariate MI

We already noted that the total MI, defined in Eq. (3.3) must always be a positive quantity. When *all* the *rv* are independent, then the corresponding correlation function $g(x_1, x_2, \ldots, x_n)$ is equal to one for all values of $(x_1, \ldots, x_n)$. Therefore, the *TI* in this case is zero. Recall that the MI between two *random variables* is always positive. This is, of course, not true for the *CI*. For example, for three random variables, see Eq. (3.19), when X_1 and X_2 are independent, then $I(X_1; X_2) = 0$. Similarly, when X_1 and X_3 are independent, we have $I(X_1; X_3) = 0$. Clearly, in this case the quantity $I(X_1; (X_2, X_3))$ does not have to be zero, even when $I(X_1; X_2) = I(X_1; X_3) = 0$. On the other hand, in this case $CI(X_1; X_2; X_3)$ would be *negative*. Intuitively, we expect that if X_1 and X_2 are independent, and also X_1 and X_3 are independent), then also X_1 would be also independent of (X_2, X_3). In other words, if having information about X_1 does not convey any information on X_2, and having information about X_1 does not convey any information on X_3, then we intuitively expect that information about X_1 will not convey any information on the joint *rv* (X_2, X_3).

To understand the origin of the negative value of *CI*, we examine Eq. (3.17), where on the right hand side the equation we have only *total correlation*. Clearly, if each of the *pair* correlations is 1, it does not follow that the *triplet* correlation will also be 1. In this case, we obtain again a negative *CI*. When there are no correlations between each pair in (3.17), we intuitively expect that there should be no correlation between the triplet, and (3.17) is negative. This case is referred to as "frustration," by Matsuda [3]. We shall see below a few examples of frustrated systems. We also note that in the literature, the conditional MI appears under different names, e.g. "*Interaction of information*" (McGill [5]), "*co-information*" (Bell [6]). Sometimes, the terms "*redundancy*" or "*synergy*" are also used for this quantity. Note also that in this book we use the definition of $CI(X_1; X_2; X_3)$ as in Eq. (3.14).

In Sect. 4.8 of Ben-Naim [1] we presented some qualitative examples of negative $CI(X_1; X_2; X_3)$. We present here another qualitative example of a case where $I(X_1; X_2) = 0$ (i.e., X_1 and X_2 are independent), but $I(X_1; X_2|X_3) > 0$, i.e. knowing X_3 would *induce* correlation between X_1 and X_2.

Consider two binding sites 1 and 2 which can bind a neutral atom, say argon. See Fig. 3.1a, each site may be either empty or occupied. If there are no interactions between the two atoms on two sites, then there is no correlation between the states of sites 1 and 2. Therefore, $I(X_1; X_2) = 0$. Now, we add an ion at site (3). This ion can polarize the two atoms on sites (1) and (2), Fig. 3.1b. In this case, the two polarized atoms can interact, and therefore the conditional MI, $I(X_1; X_2|X_3)$ will not be zero. Hence, by definition (3.8) $CI(X_1; X_2; X_3)$ will be negative. In this case it is not clear why a negative CI should be referred to as "frustration." In the next section, we shall study spin systems where the term frustration is appropriate.

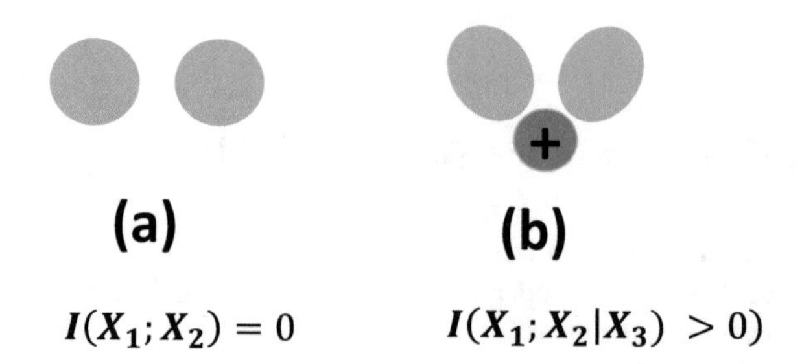

Fig. 3.1 **a** Two atoms adsorbed on two sites at a distance such that there is no interaction between the atoms. **b** A charges particle at a third site may cause polarization of the two atoms, inducing interaction between them, hence the conditional MI is positive

3.8 A Three-Spin System

In this section, we will study a *quantitative* example of three random variables for which we can calculate all the probability distributions. Hence, we can calculate all the relevant SMI, as well as the relevant mutual information.

In the first part of this section we present the method of calculating all the relevant probabilities. This is based on statistical thermodynamics. The reader does not need to understand the method we use to calculate these probabilities. Readers are urged to check that the calculated probabilities *make sense*. Once these probabilities are accepted, the reader is urged to carefully examine all the derived quantities. Later on, we shall extend the discussion for larger numbers of spins.

The system is an extension of the two-spin system we discussed in Sect. 3.4 of Ben-Naim [1]. Here, we have three spins situated at the vertices of a regular triangle as shown in Fig. 3.2.

The total number of configurations is $2^3 = 8$. Each spin can be in either the "up," or "down" state. Assigning the value of $(+1)$ to the "up" state, and (-1) to the down state, we write the *interaction energy* for any configuration of the three spins as:

$$U(x, y, z) = -J \times (x \times y + x \times z + y \times z) \tag{3.25}$$

In Eq. (3.25) $J > 0$ corresponds to a ferromagnetic behavior and $J < 0$ to anti-ferromagnetic behavior. In the study of this system we will not be interested in the physical meaning of this system, but only in the probabilities of different events, the corresponding SMI and the mutual information. All the probabilities are derived from the following equation:

$$p(x, y, z) = \frac{\exp[-\beta U(x, y, z)]}{Z_3} \tag{3.26}$$

where Z_3 is the normalization constant defined

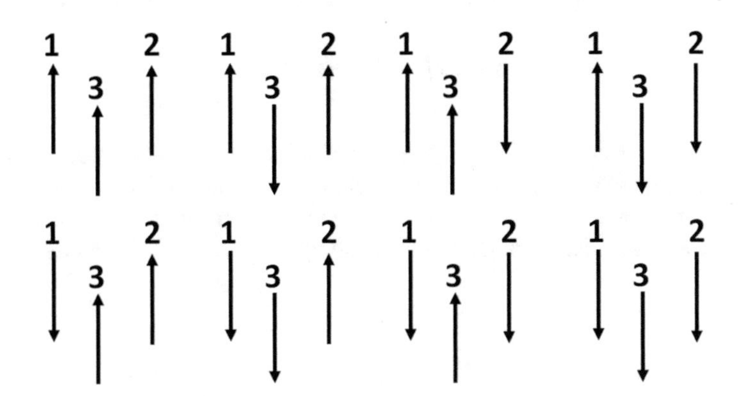

Fig. 3.2 All eight configurations of the three spins situated at the vertices of a regular triangle

$$Z_3 = \sum_x \sum_y \sum_z \exp[-\beta U(x, y, z)] \tag{3.27}$$

We take $\beta = \frac{1}{T}$ where T is the absolute temperature, and we put the Boltzmann constant $k_B = 1$. The sum in Eq. 3.27 is over all possible values of x, y, z which in our case are ± 1. Thus, $p(x, y, z)$ is the probability of finding the spin number one (X) to be in state x (up or down), spin number two (Y) to be in state y, and spin number three (Z) to be in state z. Thus, $p(x, y, z)$ is a shorthand notation for:

$$p(x, y, z) = \Pr(X = x, Y = y, Z = z) \tag{3.28}$$

Note that all the eight configurations in Fig. 3.2 are of two types; either all three spins are in the same state ("up-up" and "down-down" with interaction energy $-3J$, or one spin is in one state ("up" or "down"), and the other two in the other state ("down" or "up"). There are six such configurations in Fig. 3.2. The corresponding energy of the latter configuration is $+J$.

Thus, Z_3 may be written as:

$$Z_3 = 2\exp[+3\beta J] + 6\exp[-\beta J] \tag{3.29}$$

For $T \to 0$, and $J > 0$ the two configurations of the first type have the *higher* probability. For $T \to 0$, and $J < 0$ the six configurations of the second type have the *higher* probability. Such configurations are often referred to as being *frustrated* in the sense that not all *pairs* of spins can be "satisfied," i.e. be in opposite directions.

3.8.1 Probabilities

All the triplet probabilities are calculated from Eq. (3.26). Once we know $p(x, y, z)$, we can calculate all the marginal probabilities. The pair-probability is defined as:

$$p(x, y) = \sum_z p(x, y, z) \tag{3.30}$$

Here, the sum is over all possible values of z. Similarly $p(y)$ and $p(z)$ are defined as the marginal probabilities of $p(x, y)$. In Chap. 4 of Ben-Naim [1] we studied in great details all the probabilities in this system and their dependence on the temperature and the parameter J. Here we go directly to study the relevant SMI and the MI.

3.8.2 SMI and Conditional SMI

We start with the SMI of the single spin. In this system $H(X) = H(Y) = H(Z)$. As expected, the SMI for a single spin is always 1 (i.e., equal probability for the two states "up" and "down").

Figure 4.16 of Ben-Naim [1] shows the conditional SMI, $H(X|Y = 1, \_)$. The behavior of this quantity is similar to the case of two spin system, the lower bar "_" is to remind us that here a third spin is present, but its state is unspecified. For $J > 0$ and $T \to 0$, given $Y = 1$, the most probable configuration is all up.

The same is true for $H(X|Y = -1, \_)$. Therefore, also the average of the two, $H(X|Y, \_)$ has the same behavior as $H(X|Y = 1, \_)$. Hence, we have the equality

$$H(X|Y = 1, \_) = H(X|Y = -1, \_) = H(X|Y, \_) \tag{3.31}$$

For $J > 0$ and $T \to 0$, we have only one stable configuration (all "up"), and therefore the values of the SMI start at $H(X|Y = 1, Z = 1) = 0$.

Next, we turn to the SMI of two spins in the presence of the third one. Figure 3.3 shows $H(X, Y|Z = 1)$ for $J \geq 0$. The range of variation of the values of this SMI is between 0 and 2. Given $Z = 1$, and $J > 0$ at $T \to 0$, we have only one configuration with probability one, this is the all up configuration and the corresponding value of the SMI is zero. For very high temperatures we have equal probabilities to all the four configurations of X and Y given $Z = 1$. Therefore, the conditional SMI will tend to 2 for $T \to \infty$.

It is easy to see that the same curves are obtained for $H(X, Y|Z = -1)$, and therefore also for the average of those cases. Hence, we have the equality

$$H(X, Y|Z = 1) = H(X, Y|Z = -1) = H(X, Y|Z) \tag{3.32}$$

Next, we show in Fig. 3.4 the SMI for the three spins, as a function of T with various values of J. For $J > 0$, and $T- > 0$ there are only two stable states with

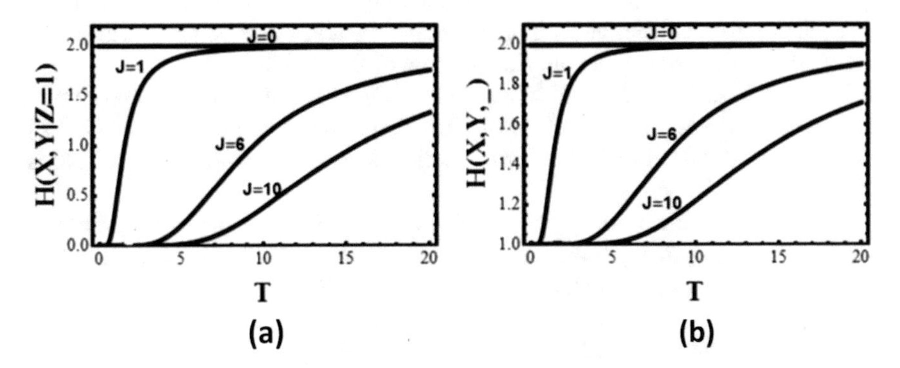

Fig. 3.3 The conditional SMI **a** H(X, Y|Z = 1) and **b** H(X,Y,_) as a function of T for three positive values of J

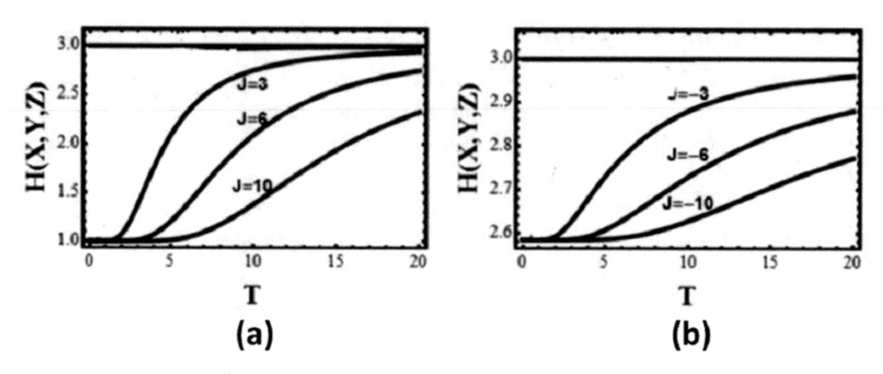

Fig. 3.4 The SMI for the three spins H(X, Y, Z) as a function of T for three **a** positive values and **b** three negative values of J

equal with equal probabilities either "all-up" or "all-down." Therefore, in this limit $H(X, Y, Z) = 1$. As the temperature increases the value of the SMI will tend to 3. The reason is that as $T \to \infty$ the effect of the temperature overcomes the interactions between the spins, and we get eight configurations having equal probabilities.

A different behavior is shown for the case $J < 0$. Here, again at $T \to \infty$ all the curves tend to 3 as shown in Fig. 3.4b. However, when $T \to 0$ the two configurations "all-up" and "all-down" will have zero probability leaving six configurations with equal probabilities. Therefore, the limiting value of $H(X, Y, Z)$ at $T \to 0$ will be $\log_2 6 \approx 2.58$.

3.8.3 *The Various Mutual Information for Three-Spin Systems*

In Chap. 4 of Ben-Naim [1] we calculated all the relevant correlations for a three-spin system and showed their behavior as a function of T for various values of J. Here, we shall compare a linear and a triangle arrangement of the three spins. This study is important in understanding the phenomenon of frustration.

Figure 3.5 shows the mutual information (MI) of X and Y in the presence of a third spin (but in an unspecified state). This should be compared with $I^{(2)}(X : Y)$ in Fig. 3.9 of Ben-Naim [1]. The latter notation is used to remind us that this MI is calculated in a two-spin system.

Likewise, we could have written $I^{(3)}(X; Y)$ for the MI between X and Y calculated in the three-spin system. However, in order to avoid confusion with the MI for the three spins, see below, we use the notation $I(X; Y, \_)$, where the lower bar "_" stands for the presence of a third spin, the state of which is unspecified.

In Fig. 3.5a, we show the MI as a function of T for $J > 0$, we see that the general behavior of $I(X; Y, \_)$ is similar to that of $I^{(2)}(X; Y)$ in Fig. 3.9 of Ben-Naim [1]. When $J = 0$, there is no correlation between the two spins, hence $I(X; Y, \_) = 0$.

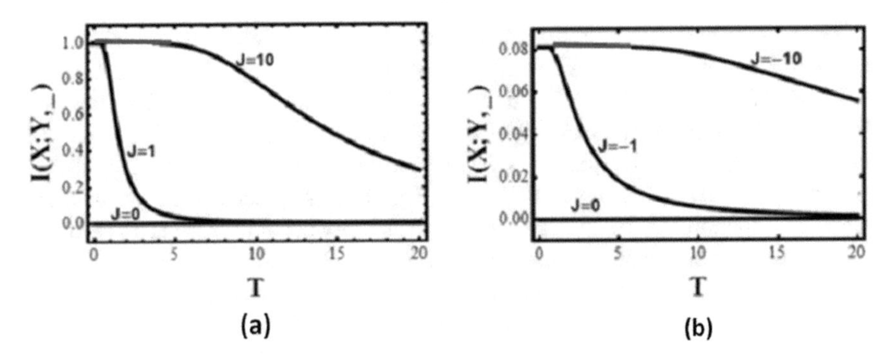

Fig. 3.5 The mutual information in the presence of a third spin as a function of T for **a** two positive and **b** two negative values of J

On the other hand, turning on the interaction ($J > 0$) induces a correlation between X and Y. When $T \to 0$, the MI is one, which means that knowing X provides full information on Y, and vice versa. Unlike the case of two-spin systems where we observed the same behavior for $J > 0$ and $J < 0$, here in the presence of a third spin, the behavior of $I(X; Y, \_)$ is very different in the case $J < 0$, Fig. 3.5b.

First, note that in the range of values in Fig. 3.5b we see that when we have strong interaction $J = -10$, the maximum MI is 0.08. This means that strong interaction ($J \approx -10$) does not induce large MI; knowing X provides little information on Y and vice-versa.

The reader should pause and think about the difference in behavior of the MI in Fig. 3.5a, b. Now, consider the case of very low temperature $T \to 0$. When $J > 0$, knowing $X = 1$ induces two effects on Y, the *direct* interaction between X and Y will cause Y to favor "up." In addition, X will cause also Z to prefer "up," and Z, in turn will cause Y to favor "up." Thus, we see that both the direct and the indirect interactions contribute to *positive* correlation between X and Y.

Next, consider the case of $X = -1$ ("down"). This causes Y to prefer "down." It also causes Z to favor "down," which in turn causes Y to be "down." We see again that both the direct and the indirect interactions produce positive correlation between X and Y. This is the reason for the very large value of $I(X; Y, \_)$ in Fig. 3.5a. ($J > 0$, and $T \to 0$).

Now, let us examine the case of $J < 0$ and $T \to 0$ knowing that $X = 1$ ("up") causes Y to prefer "down." In addition, $X = 1$ causes Z to favor "down," which pushes Y towards "up." We see that the direct and the indirect interactions induces opposite effects. The direct interaction induces a negative correlation, whereas the indirect interaction induces positive correlation.

Next, still with $J < 0$, suppose we know that $X = -1$ ("down"). The direct interaction causes Y to prefer "up" (negative correlation). In addition, $X = -1$ causes Z to favor "up," and therefore Z pushes Y toward "down." Therefore, the direct interaction produces positive correlations.

Thus, in the presence of Z, we see that knowing X provides much information on Y, when $J > 0$, but very little information when $J < 0$.

Figure 4.27 of Chap. 4, Ben-Naim [1] shows the values of $2H(X|Z)$ (blue line) and $H(X, Y|Z)$ (red line) for the choice of $J = 10$. We see that both curves increase with temperature but at slightly different rates. We can also see that the difference between $2H(X|Z)$ and $H(X, Y|Z)$ is initially zero, then becomes increasingly positive and at some temperature starts to decrease, and eventually becomes zero at $T \to \infty$.

Note that the range of variation of $(X; Y|Z)$, in Fig. 3.6 is quite small. We can say that the condition Z has a negligible effect on the MI between X and Y.

Figure 3.7 shows the values of the conditional MI defined by:

$$CI(X; Y; Z) = I(X; Y) - I(X; Y|Z) \tag{3.33}$$

For $J > 0$, we saw that $I(X; Y|Z)$ is negligibly small. Therefore, the *CI* in Eq. (3.33) is dominated by $I(X; Y)$ which is positive. This is essentially the same as $I(X; Y, \_)$. On the other hand, for $J < 0$, the mutual correlation $I(X; Y)$ which is the same as

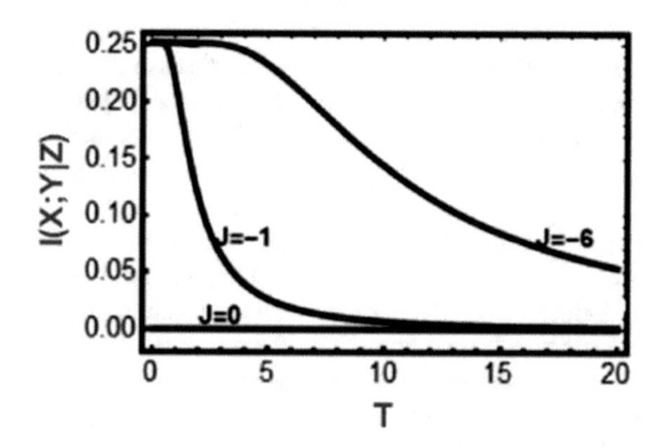

Fig. 3.6 The conditional MI, H(X, Y|Z), as a function of *T* for two negative values of *J*

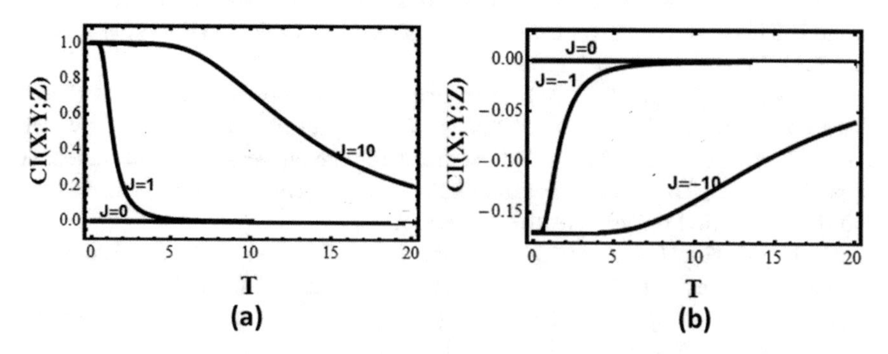

Fig. 3.7 The conditional triplet MI as a function of *T* for **a** two positive and **b** two negative values of *J*

$I(X; Y, \_)$, is negligibly small. In this case, the quantity $I(X; Y|Z)$ dominates the *CI*, which is *negative*. The fact that *CI* is negative is attributed to the *frustration* in the three-spin system.

Finally, we show in Fig. 3.8 the total MI between *X*, *Y* and *Z* defined by

$$TI(X; Y; Z) = \sum_{x,y,z} p(x, y, z) \log[g(x, y, z)] \tag{3.34}$$

These quantities must always be positive (see Sect. 3.4). We see that this MI is large and positive for $J > 0$, but relatively small for $J < 0$. The reason is that the triplet correlations in the case $J > 0$ are much larger than in the case $J < 0$. For more details, see Chap. 4, Ben-Naim [1].

The linear configuration

Figure 3.9 shows all eight configurations of the linear arrangement of the three spins. When $J > 0$, the "all-up" or "all down" configuration will have the lowest energy of interaction.

Figure 3.10 shows the three probabilities for $J = 1, T = 1$:

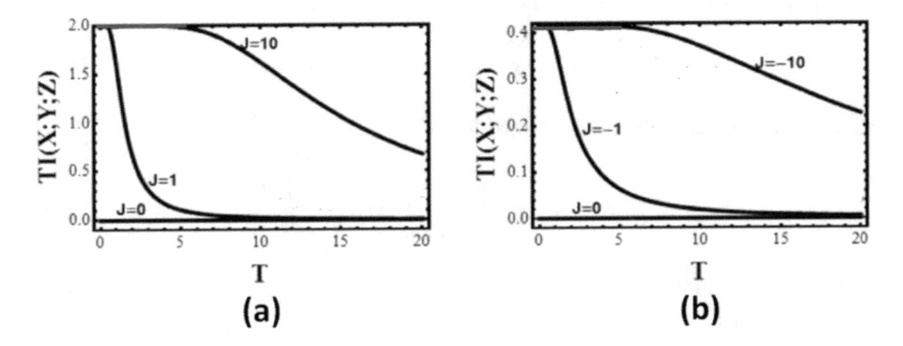

Fig. 3.8 The total MI as a function of *T* for **a** two positive and **b** two negative values of *J*

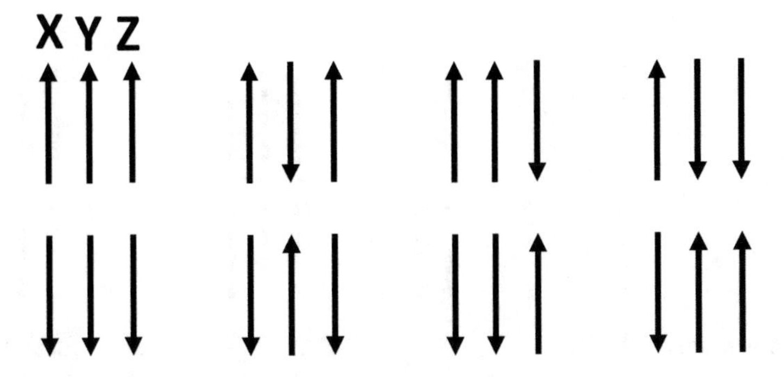

Fig. 3.9 All eight configurations of the three linear spin-system

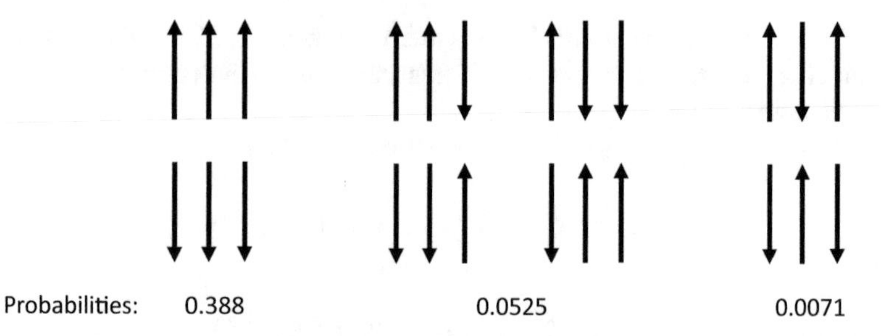

Fig. 3.10 The probabilities of the three "energy levels" for $J = 1$

2 configurations all "up" or "all down" with probability: Pr = 0.388.
4 configurations with probability: Pr = 0.0525.
2 configurations with probability: Pr = 0.0071.
Figure 3.11, shows the probabilities for $J = -1, T = 0$:
2 configurations Pr = 0.388
4 configurations Pr = 0.0525
2 configurations Pr = 0.0071
The interaction energy is given by:

$$U(x, y, z) = -J(x_1, x_2) = -J(x_1 \times x_2 + x_2 \times x_3) \tag{3.35}$$

and the probabilities are calculated from:

$$p(x, y, z) = \frac{\exp[-\beta U(x, y, z)]}{\sum \exp[-\beta U(x, y, z)]} \tag{3.36}$$

where $\beta = \frac{1}{T}$, and the sum is over all possible values of x, y, z (here, $+1$ and -1 for "up" and "down", respectively).

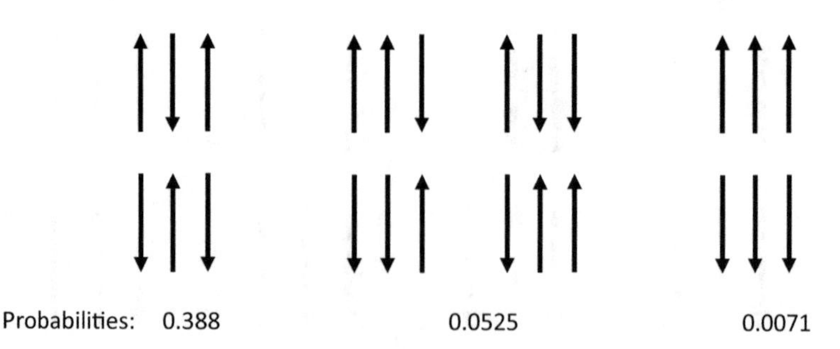

Fig. 3.11 The probabilities of the three "energy levels" for $J = -1$

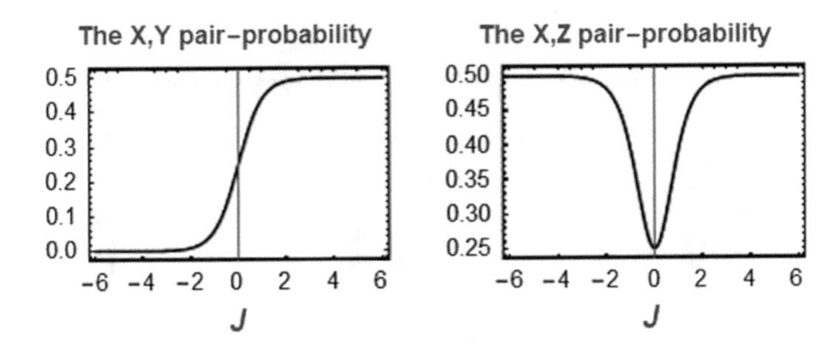

Fig. 3.12 The pair-probabilities for nearest neighbors, either *X, Y or* Y, Z, and next nearest neighbors *X, Z,* as a function of *J for a fixed temperature*

As we have discussed earlier, all the required probabilities may be derived from the triplet probability given in Eq. (3.36). As in the case of the triangle configuration, here, also the singlet probability is: $p(1) = p(-1) = 0.5$. Unlike the triangle case where all pairs are equivalent, here the pair probabilities are different for the pair X, Y and the pair X, Z.

Figure 3.12 shows the different behavior of $p(x = 1, y = 1)$ and $p(x = 1, z = 1)$ as a function of J for a fixed temperature $T = 1$. In both cases, when $J = 0$ the two spins are independent. Therefore, the configuration "up-up" has probability 0.25 (i.e., all the configurations of the pair of spins are equally probable). For very large and positive J, say $J = 6$, the dominant configurations are either "up-up" or "down-down," which are equally probable. However, when J is large and negative the probability of the configuration $x = 1$ and $y = 1$ (i.e. up-up for the pair X and Y) will tend to zero. This is the case because whatever the configuration of Z is, the configuration of the "up-up" or "down-down" will not be favorable for the case $J = -6$. The probability of $x = 1, y = 1$, however will be 0.5 for the case $J = -6$ since in this case either the "up-down-up" or "down-up-down" configuration will become dominant. Hence, the configuration of the pair $X, Z, x = 1, z = 1$ will have probability 0.5.

Next, we shall discuss the various SMI in this system. As expected, the singlet SMI is equal to one for any T and J, i.e.:

$$H(X) = H(Y) = H(Z) = 1 \tag{3.37}$$

The pair-SMI also behaves as expected; for $J = 0$ we have

$$H(X, Y) = H(X, Z) = H(Y, Z) = 2 \tag{3.38}$$

(all four configurations are equally probable).

For large J (either positive or negative) the three-pair SMI will tend to one (i.e. only two configurations are equally probable). Note however, that the slope of the two curves are slightly different for the X, Y and the Y, Z pairs.

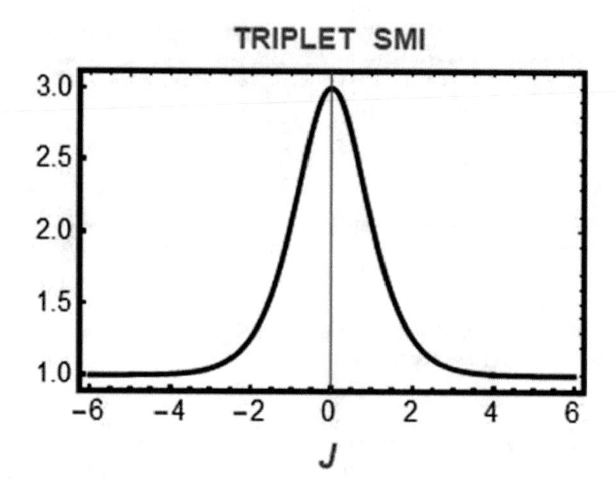

Fig. 3.13 The triplet SMI for the linear system as a function of J for a fixed temperature

The triplet SMI starts at value of $H(X, Y, Z) = 3$ (i.e. all eight configurations are equally probable) at $J = 0$, and ted to one for large J (either positive or negative), Fig. 3.13.

This is quite different from the behavior of $H(X, Y, Z)$ for the triangle case where the limit of this SMI for large negative J is $\log_2 6 = 2.58$, i.e. only six out of the eight configurations are equally probable at this limit (for details, see Sect. 4.9, Ben-Naim [1]).

Next, we shall discuss a few of the MI in this system. As we noted in Ben-Naim [1], there are several ways of defining multivariate MI. The simplest MI is the one we referred to as the total MI and denoted by $TI(X; Y; Z)$. Figure 3.14 shows the values of $TI(X, Y, Z)$ for the linear system as a function of T for different values of J. As can be seen the general behavior of $TI(X; Y; Z)$ as a function of T is quite similar for positive and negative values of J.

Finally, we show in Fig. 3.15 the conditional MI denoted $CI(X; Y; Z)$.The most interesting in these two figures is that $CI(X; Y; Z)$ is the same for both positive and negative values of J. This is in sharp contrast to the case of the triangle configuration shown in Fig. 4.30 of Ben-Naim [1]. In the triangle case the values of CI for negative

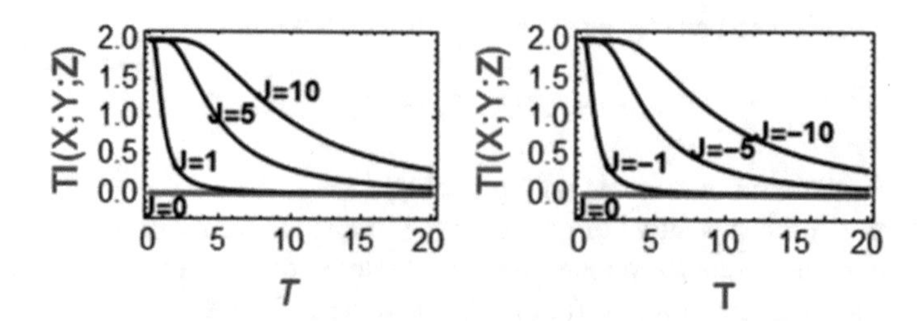

Fig. 3.14 The total MI as a function of temperature for different values of J

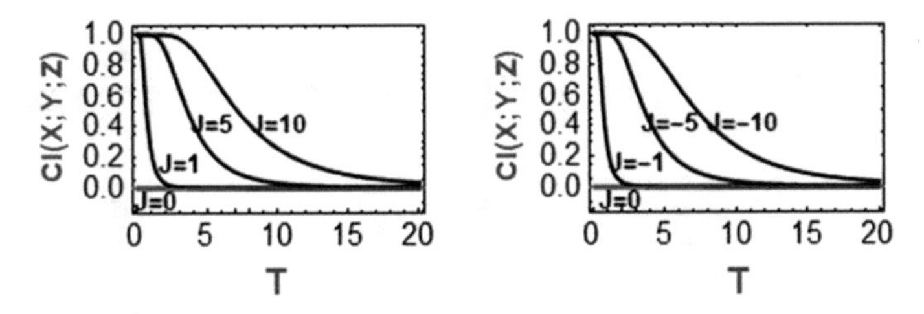

Fig. 3.15 The conditional MI as a function of temperature for different values of J

J, are negative. This finding was explained by Matsuda as due to the frustration in the triangle case. In the linear case, on the other hand, the values of CI are all positive and are the same for positive and negative values of J. Figure 3.15 shows the total and the conditional MI for the linear system of three spins as a function of βJ.

3.8.4 The Three-Spin System with External Field

In this section we study the effect of an external field on the MI between the three spins in the triangular configuration. Since we did not specify the type of the dipoles we assume here that the interactions between any pair of spins is as in Eq. 3.39, below. In addition, we add an additional field (f), which favors the "up" orientation of the spin. Thus, the total energy of the system is given by:

$$\begin{aligned} U(x_1, x_2, x_3) = &-J(x_1 \times x_2 + x_1 \times x_3 + x_2 \times x_2) \\ &- f(x_1 + x_2 + x_3) \end{aligned} \tag{3.39}$$

All the probabilities for this system are calculated as before from Eq. 3.36. For the case $f = 0$ (i.e. no external field) we have for $\beta J = 1$, two energy levels; "all up" and "all down" with probability 0.474, and four other higher energy levels with probability of 0.087. The total MI, (TI) and the conditional MI (CI) are shown as a function of βJ in Fig. 3.16. (βJ is simply $\frac{J}{k_B T}$ with $k_B = 1$).

Note that in this case we have the limiting values of CI and TI as:

$$\begin{aligned} \lim_{\beta J \to -\infty} CI = -0.17, \quad \lim_{\beta J \to \infty} CI = 1 \\ \lim_{\beta J \to -\infty} TI = 0.4, \quad \lim_{\beta J \to \infty} TI = 1 \end{aligned} \tag{3.40}$$

Turning on a small external field, $\beta f = 0.1$ has a negligible effect on either CI or TI, however there are now four, rather than two energy levels with corresponding probabilities:

For "all up" configuration $\Pr = 0.613$, but for the "all down" configuration $\Pr = 0.336$. Also, the six configurations, which had equal probabilities in the case

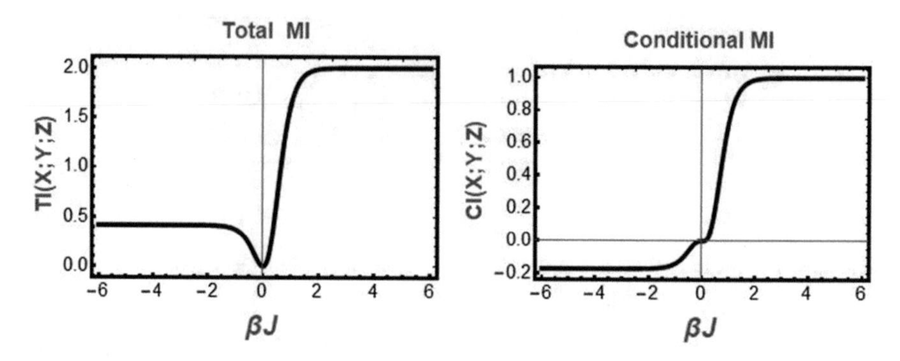

Fig. 3.16 The total and the conditional MI for the three spins in a triangle arrangement as a function of $\boldsymbol{\beta J}$, without external field

$\beta J = 0$ are now split into two groups:

$$p(1, -1, 1) = p(1, 1, -1) = p(-1, 1, 1) = 0.0092$$

$$p(1, -1, -1) = p(-1, -1, 1) = p(-1, 1, -1) = 0.0075$$

All these changes in the probabilities are understandable as the effect of the external field favors the "up" orientation of each of the spins.

Next, we increase the external field to $\beta f = 0.5$. We see quite a change in both CI and TI, Fig. 3.17. The limiting values are:

$$\lim_{\beta J \to -\infty} CI = -0.22, \lim_{\beta J \to \infty} CI = 0.275$$

$$\lim_{\beta J \to -\infty} TI = -0.52, \lim_{\beta J \to \infty} TI = 0.55$$

Figure 3.18 shows CI and TI as a function of βJ for $\beta f = 1$. Here, we see a more dramatic effect. The limiting values are now:

$$\lim_{\beta J \to -\infty} CI = -0.31, \lim_{\beta J \to \infty} CI = 0.025$$

$$\lim_{\beta J \to -\infty} TI = -0.75, \lim_{\beta J \to \infty} TI = 0.05$$

Finally, we show in Fig. 3.19 the values of CI and TI for the case $\beta f = 2$.

Note that as we increase βf the limiting value of CI at large and negative βJ becomes more and more negative. For the case $\beta f = 2$ we have $\lim_{\beta J \to -\infty} CI = -0.4$, and for $\beta f = 10$ we have $\lim_{\beta J \to -\infty} CI = -0.415$. All this limiting values can be understood by calculating the relevant SMIs.

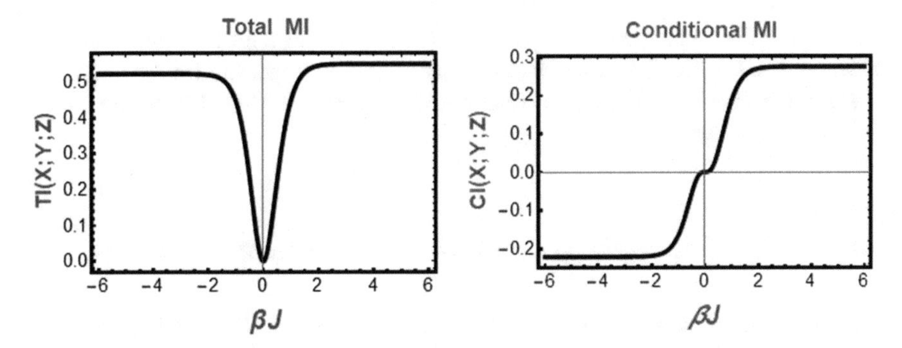

Fig. 3.17 The total and the conditional MI for the three spins in a triangle arrangement as a function of βJ. With external field $\beta f = 0.5$

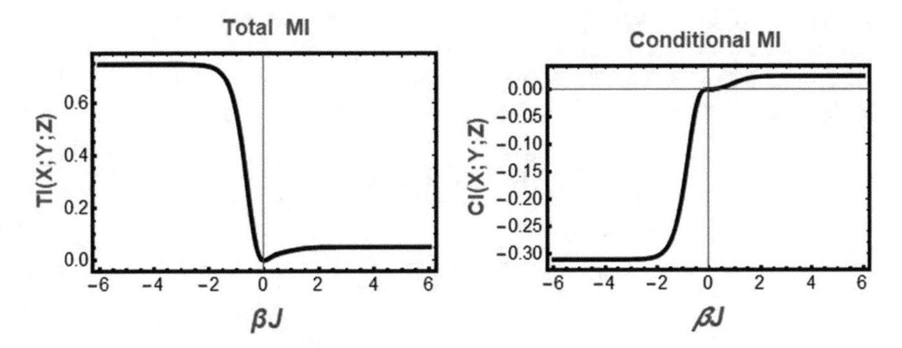

Fig. 3.18 The total and the conditional MI for the three spins in a triangle arrangement as a function of βJ. With external field $\beta f = 1$

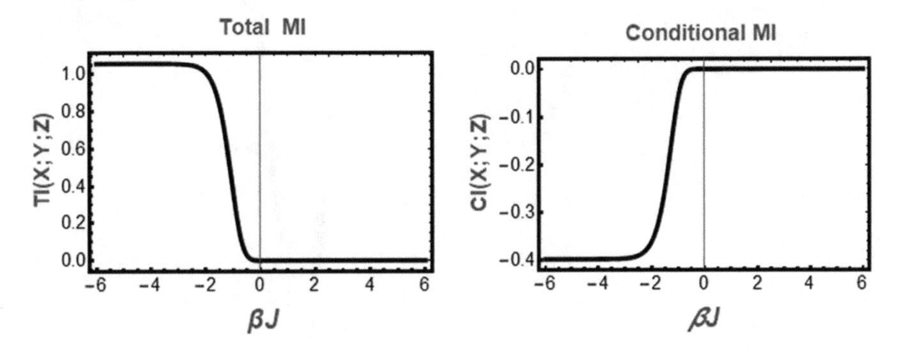

Fig. 3.19 The total and the conditional MI for the three spins in a triangle arrangement as a function of βJ. With external field $\beta f = 2$

We note that negative values of CI were interpreted as a measure of frustration. However, it is difficult to argue that the larger βf, the larger the frustration in this system. We shall say more on this relationship between the CI and frustration in the following sections.

3.8.5 The Three-Spin System with Different Geometries

We discuss here another example of a three-spin system but with changing the geometry of the configuration of the triangle. We fix the distance between the two spins X_1 and X_2, and change the distance of the third spin X_3 as shown in Fig. 3.20.

We choose the distance between X_1 and X_2 as unity ($d = 1$). The distance between X_3 and X_1 will be the same as the distance between X_3 and X_2 and will be denoted d (in units of the distance between X_1 and X_2).

Figure 3.21 shows the total and the conditional MI for the distance $d = 2$. The limiting values of CI and TI in this case are:

$$\lim_{\beta J \to -\infty} CI \approx 0 \quad \lim_{\beta J \to \infty} CI = 1$$

$$\lim_{\beta J \to -\infty} TI = 1 \quad \lim_{\beta J \to \infty} TI = 2$$

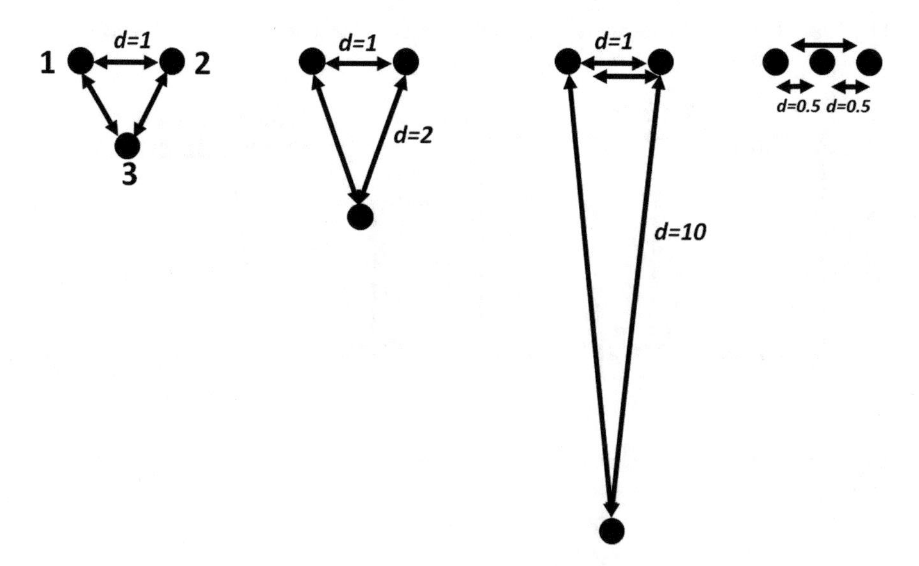

Fig. 3.20 The four geometrics of the triangle arrangement studied in Sect. 3.8; the perfect triangle with edge $d = 1$ (left), $d = 2$ $d = 10$ and $d = 0.5$ (right)

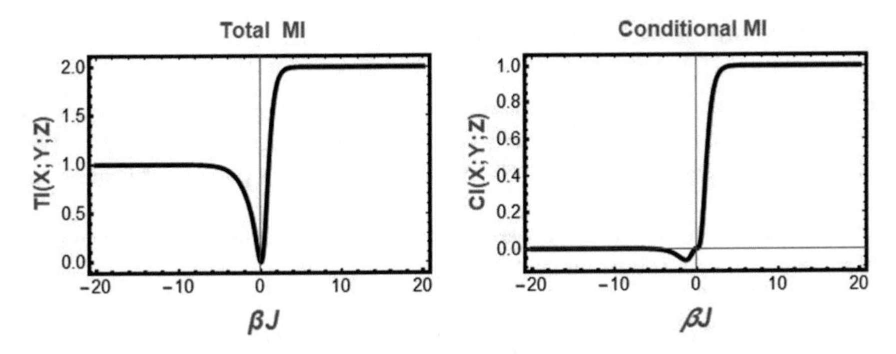

Fig. 3.21 The total and the conditional MI for the three spins in a triangle arrangement as a function of βJ. With $d = 2$

Note that for small and negative values of βJ the CI is negative and it approaches zero for $\beta J \to -\infty$.

Figure 3.22 shows the CI and TI for $d = 10$. At this distance the interaction between X_3 and either X_1 or X_2 is almost zero. We do not expect any frustration in this system and we see that the CI is almost zero for the entire range of negative βJ. The limiting values of both CI and TI at $\beta J \to \infty$ are the same as in the case of $d = 2$, i.e. they are 1 and 2, respectively.

Finally, we bring X_3 to a distance $d = 0.5$, which means we have a linear arrangement of the three spins see Fig. 3.20. Note however, that unlike the linear case discussed in Sect. 3.8.3, where we had only nearest-neighbors interactions (between X_1 and X_2, and between X_2 and X_3), here we have interactions between the three pairs of spins. We leave this case as an exercise. Here, we note that the limiting values of CI at both limits $\beta J \pm \infty$ is one, and for TI it is 2. One interesting aspect of CI is that for small negative $\beta J \approx -0.5$ we have a small negative value of CI. The reader is urged to ponder on the question whether we can say that this negative CI is due to "frustration," and if it is why this effect disappears when $\beta J \to -\infty$.

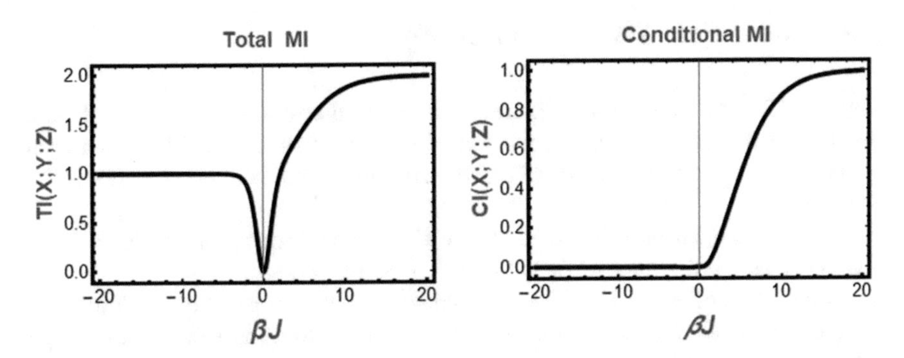

Fig. 3.22 The total and the conditional MI for the three spins in a triangle arrangement as a function of βJ. With $d = 10$

3.9 Systems with Four Interacting Spins

In this and in the next few sections we study the MI among larger number of spins. As we have noted in Chap. 4 of Ben-Naim [1], there are several possible generalizations of the MI. In Chap. 4 of Ben-Naim [1] we discussed mainly two generalizations of the MI.

The first was referred to as the *total* MI denoted TI. This is defined for any number of random variables (or experiments) by:

$$TI(X_1; \ldots; X_n) = \sum p(x_1, \ldots x_n) \log g(x_1, \ldots, x_n) \tag{3.41}$$

where $p(x_1, \ldots, x_n)$ is the probability of finding the event $\{X_1 = x_1, X_2 = x_2, \ldots X_n = x_n)$ and $g(x_1, \ldots, x_n)$ is the correlation function, defined by:

$$g(x_1, \ldots, x_n) = \frac{P(x_1, \ldots, x_n)}{\prod_{i=1}^{n} P(x_i)} \tag{3.42}$$

The second generalization was referred to as the conditional MI, and denoted CI. There are several possible definitions of CI. Here, we use the one chosen by Matsuda [3]. This is not the most informative form of CI, however, we use this particular one to compare our results with Matsuda's results. This definition is:

$$CI(X_1; \ldots; X_n) = \sum_{k=1}^{n} (-1)^{k-1} \sum_{(i_1, \ldots, i_k)} H(X_1, \ldots X_n) \tag{3.43}$$

Here, the sum on the right hand side of (3.43) is over-all possible sets of indices $(i_1, i_2, \ldots, i_k)$, with : $1 < i_1 < i_2 < \ldots i_k < k$.

As we have noted several times, this quantity may be negative.

3.9.1 *Four-Spin Systems; Perfect Square*

Figure 3.23 shows three possible arrangements of the four spins system: (a) is a regular square with equal edges. (b) is a parallelogram in which the distance between 1 and 3 is the same between 1 and 3 and (c) a rectangle with two short and two twice longer edges.

For the arrangement of Fig. 3.20a the total number of configurations is $2^4 = 16$. Again, we assign the value of $(+1)$ to the "up" and (-1)…

In Sect. 4.10 of Ben-Naim [1] we presented some details about the various SMI, the pair and triplet MI, etc. Here, we proceed directly to discuss only the total MI and the conditional MI.

Figure 3.24 shows TI and CI for the square arrangement. For $\beta J = 0$ (either $J = 0$ or $T \to \infty$) there is no correlation between the spins. Hence, both TI and CI

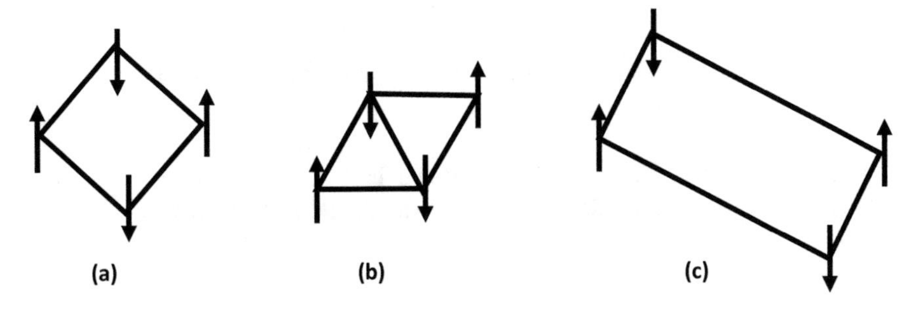

Fig. 3.23 The three geometries of the four spin systems. **a** A perfect square, **b** A parallelogram, i.e. a square with one additional interaction, and **c** a rectangle

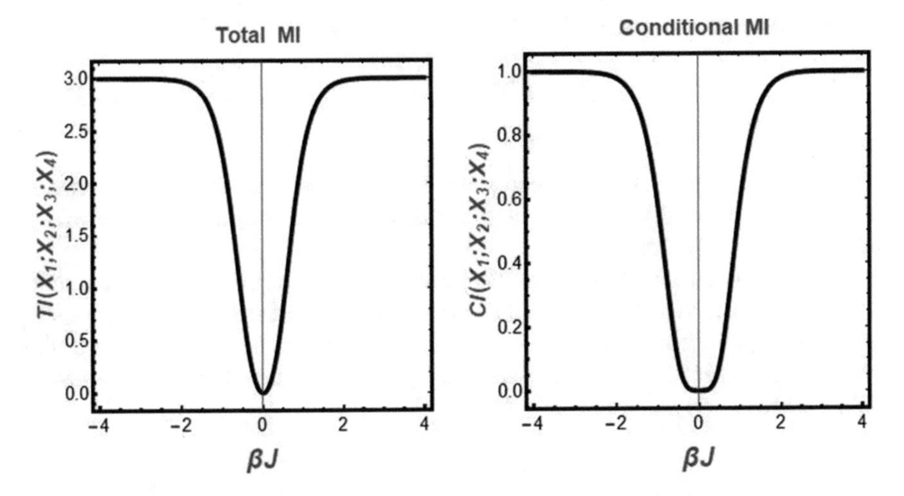

Fig. 3.24 The total and the conditional MI for the four spins in a perfect square arrangement as a function of βJ

are zero. For larger βJ (either positive or negative) the behavior is similar for both TI and CI.

For the TI, we have in this limit

$$\begin{aligned} TI(X_1; X_2; X_3; X_4) &= -H(X_1, X_2, X_3, X_4) + 4H(X_1) \\ &= -1 + 4 = 3 \end{aligned} \tag{3.44}$$

The reason for this result is clear for very strong interactions there are only two configurations for the four spins with equal probabilities. Therefore, the SMI for the four spins is one, as well as for each individual spin.

Unlike the case of three spins for which the behavior of CI is different for positive and negative βJ, here, we have the same behavior for both $\beta J \to \pm\infty$. The actual value is $\log_2 2 = 1$, which may be calculated from equation:

$$\begin{aligned} CI(X_1; X_2; X_3; X_4) &= 4H(X_1) - 4H(X_1, X_2) \\ &- 2H(X_1, X_3) + 4H(X_1, X_2, X_3) - H(X_1, X_2, X_3, X_4) \\ &\to 4 - 4 - 2 + 4 - 1 = 1 \end{aligned} \tag{3.45}$$

Note that unlike the case of triangle we do not have frustration in this case. As we noted earlier Matsuda attributed the negative value of the CI to the frustration effect.

3.9.2 *The Parallelogram Arrangement*

This arrangement is shown in Fig. 2.23b. Note that in this case we add one more interaction between 1 and 3. Thus, the potential energy is:

$$U(x_1, x_2, x_3, x_4) = -J(x_1 \times x_2 + x_2 \times x_3 + x_3 \times x_4 + x_4 \times x_1 + x_1 \times x_3) \tag{3.46}$$

Note that the interaction between 1 and 3, having the same strength as between all other nearest neighbors' pairs. In this case we do have frustration but the values of CI is not negative for $\beta J < 0$.

We show in Fig. 3.25 the total and the conditional MI. The most surprising finding here is that the curves of both TI and CI are asymmetric with respect to positive and negative βJ, all the limiting values of CI and TI are the same as in the case of the square. This is surprising as there is frustration in this system though the values of CI are positive everywhere.

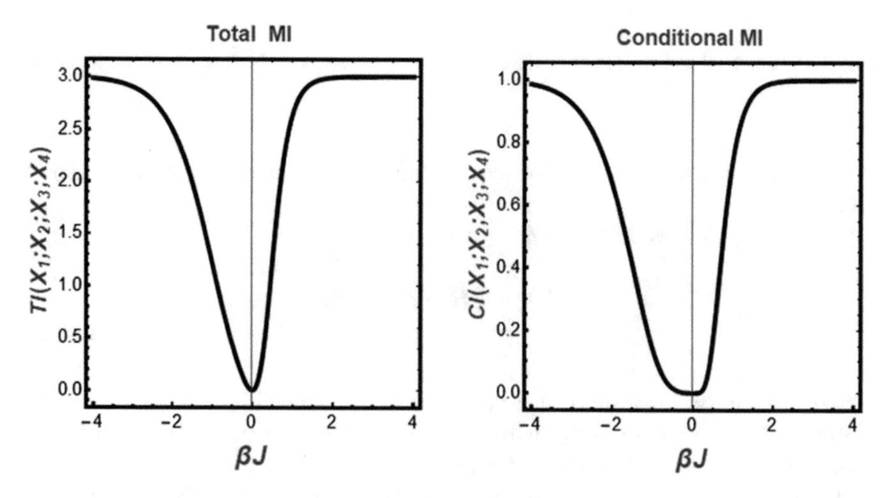

Fig. 3.25 The total and the conditional MI for the four spins in a parallelogram arrangement as a function of βJ

The curves for both TI and CI behave slightly different for positive and negative values of βJ. However, the limiting values of TI and CI is as before, namely:

$$\lim_{\beta J \to \pm\infty} TI = 3, \quad \lim_{\beta J \to \pm\infty} CI = 1$$

The fact that CI is not negative for the negative βJ casts some doubt as to the validity of Matsuda's claim that CI could serve as a measure of the frustration effect. Matsuda does not explain why we obtain a positive value of CI in spite of the existence of frustration in this system. Instead, he explains:

> Thus, the behavior of the system at finite temperature is a consequence of the frustration and the stabilization effects as well as thermal fluctuation.

Obviously, this does not explain why the "stabilization effect" overbalances the frustration effect.

3.9.3 The Rectangular Configurations

In these configurations we have two pairs at distance one (1, 4 and 2, 3) and two pairs at distance two (1, 2 and 3, 4), Fig. 3.23c. The potential energy for this case is:

$$U(x_1, x_2, x_3, x_4) = -J\left[\frac{1}{2}x_1 \times x_2 + x_2 \times x_3 + \frac{1}{2}x_3 \times x_4 + \frac{1}{2}x_4 \times x_1\right] \quad (3.47)$$

Figure 3.26 shows the total and the conditional MI for this case. Here, again the general behavior of the TI and CI is the same at both $\beta J \to \pm\infty$. The forms of the curves for this case, as is shown in Fig. 3.25 are slightly different from the case of a perfect square shown in Fig. 3.24, but the limiting behavior is the same, i.e.,

$$\lim_{\beta J \to \pm\infty} TI = 3, \quad \lim_{\beta J \to \pm\infty} CI = 1$$

Clearly, as in the case of a perfect square there is no frustration effect in the rectangular case.

3.10 Five-Spin Systems

Next, we examine the case of five spins with different arrangements. For the regular pentagon, Fig. 3.27, with only nearest neighbor interactions, we have the potential function:

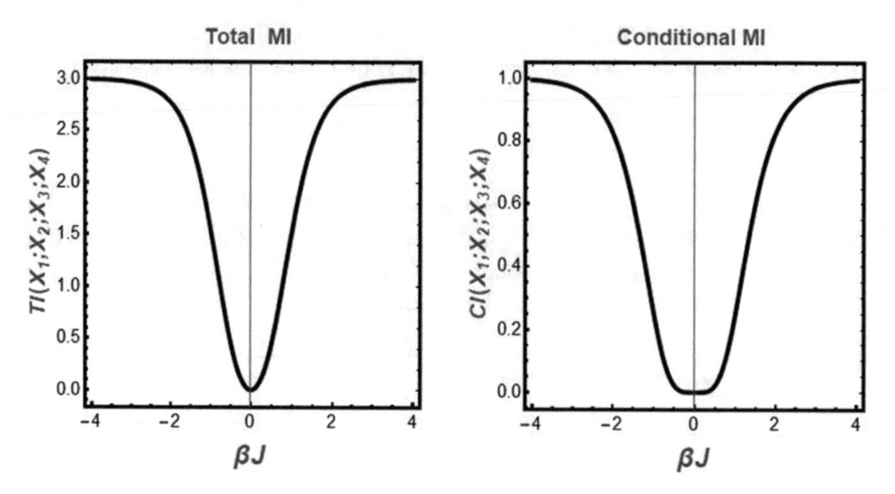

Fig. 3.26 The total and the conditional MI for the four spins in a rectangle arrangement as a function of βJ

$$U(x_1, x_2, x_3, x_4, x_5) = -J(x_1 \times x_2 + x_2 \times x_3 + x_3 \times x_4 + x_4 \times x_5 + x_5 \times x_1) \quad (3.48)$$

Unlike the perfect square case, in the regular pentagon arrangement we do have frustration effect. See Fig. 3.27.

Figure 3.28 shows the total and the conditional MI for this system. Note that in the limit of $\beta J \to \infty$, the values are as expected, i.e.

$$\lim_{\beta J \to \infty} TI = 4, \quad \lim_{\beta J \to -\infty} = 1$$

The more interesting values are in the limit of large negative βJ. For the TI, the limiting value is 1.678. The limiting value of CI is negative. As we noted earlier this is attributed to the frustration effect in this system.

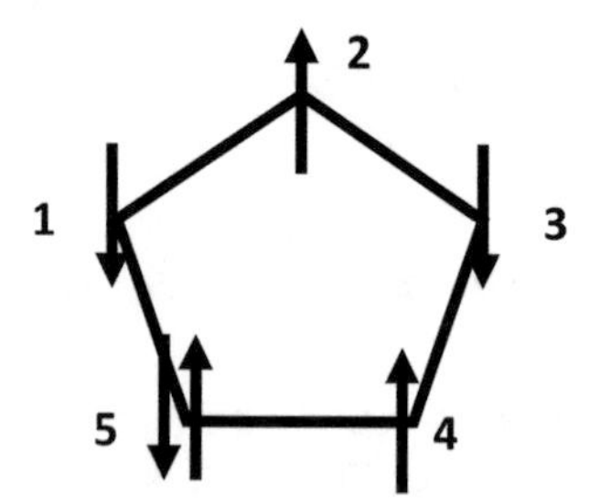

Fig. 3.27 The five-spin systems on a regular pentagon. Note that there is frustration in this system

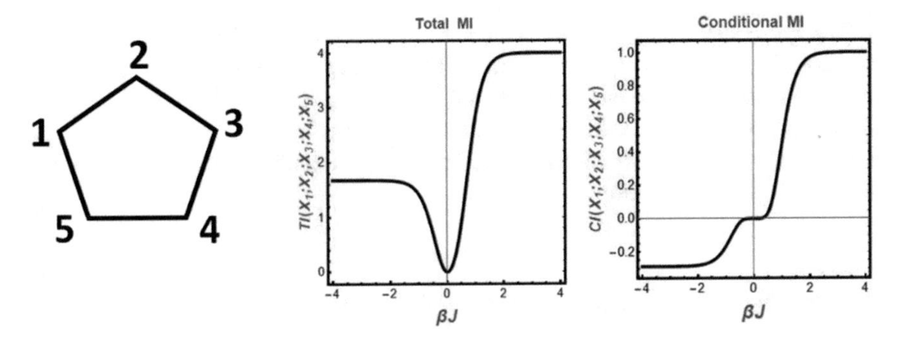

Fig. 3.28 The total and the conditional MI for the five- spins in a regular pentagon arrangement as a function of βJ

3.10.1 Pentagon with One Additional Interaction

This configuration may be viewed as a square to which we added a triangle, Fig. 3.29. It has five nearest neighbors, and in addition it has one more next-nearest neighbor between 2 and 4.

$$U(x_1, x_2, x_3, x_4, x_5) = -J(x_1 \times x_2 + x_2 \times x_3 + x_3 \times x_4 + x_4 \times x_5 + x_5 \times x_1 + x_2 \times x_4) \tag{3.49}$$

Figure 3.29 shows the total and the conditional MI. The total MI behave as expected, i.e.

$$\lim_{\beta J \to \infty} TI = 4, \quad \lim_{\beta J \to -\infty} TI = 3 \tag{3.50}$$

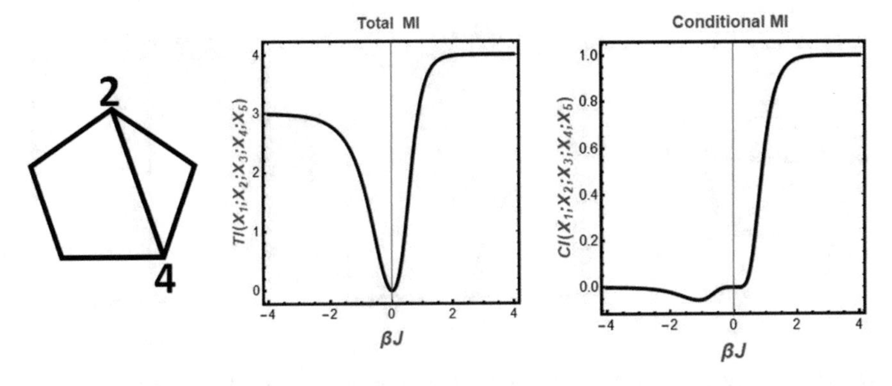

Fig. 3.29 The total and the conditional MI for the five-spins with one additional interaction, as a function of βJ

The more surprising behavior is that of CI. The limit of CI for $\beta J \to \infty$ is as expected. However, for negative values of βJ we find a strange behavior. Note that in the regular pentagon we have a negative CI for large negative βJ. This is attributed to frustration. Here, we also have frustration effect, yet in the limit of very large and negative βJ, the CI tends to zero. This finding casts doubt on the association of a negative CI with frustration.

3.10.2 Pentagon with Two Additional Interactions

This case may be viewed as three triangles combined to form a pentagon, having seven pairs of interactions. Figure 3.30.

The potential function is:

$$\begin{aligned} U(x_1, x_2, x_3, x_4, x_5) = -J(&x_1 \times x_2 + x_2 \times x_3 + x_3 \times x_4 \\ &+ x_4 \times x_5 + x_5 \times x_1 + x_1 \times x_3 + x_1 \times x_4) \end{aligned} \tag{3.51}$$

Figure 3.30 shows the total MI. In the limit $\beta J \to \infty$ the value of TI is 4 as expected, and equal to the other cases of five spins. However, in the limit $\beta J \to -\infty$ the value of TI is 2, different from the previous cases. This is a result of 8 equally probabilistic states in this limit, hence:

$$\begin{aligned} TI(X_1, X_2, X_3, X_4, X_5) &= -H(X_1, X_2, X_3, X_4, X_5) + 5H(X_1) \\ &= -3 + 5 = 2 \end{aligned} \tag{3.52}$$

Most interesting is the behavior of CI. In the limit $\beta J \to \infty$ its value is 1, as in previous cases. However, the limit of $\beta J \to -\infty$ is negative. It is similar to the regular pentagon, but different from the pentagon with one additional interaction.

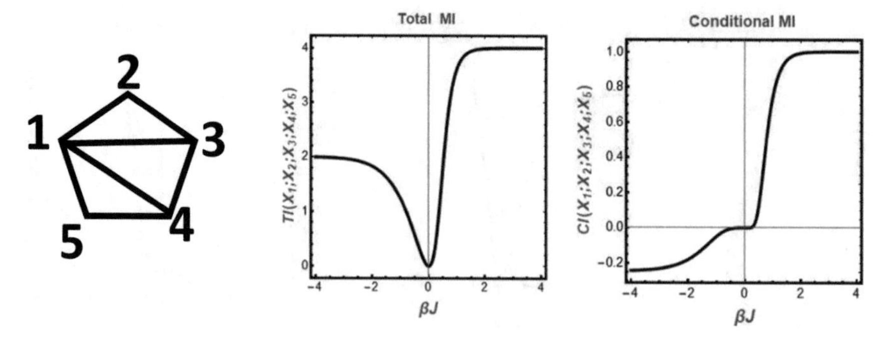

Fig. 3.30 The total and the conditional MI for the five-spins with two additional interactions, as a function of βJ

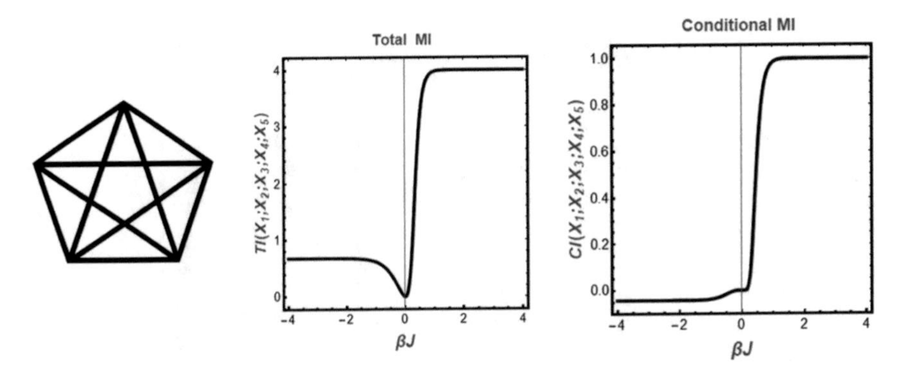

Fig. 3.31 The total and the conditional MI for the five-spins with all pair interactions, as a function of βJ

3.10.3 Pentagon with All Pair Interactions

In this case we assume that all pairs interact. There are altogether 10 pairs; (1, 2), (1, 3), (1, 4), (1, 5), (2, 3), (2, 4), (2, 5), (3, 4), (3, 5) and (4, 5), Fig. 3.31. The corresponding potential for this case is:

$$\begin{aligned} U(x_1, x_2, x_3, x_4, x_5) = &-J(x_1 \times x_2 + x_2 \times x_3 + x_3 \times x_4 + x_4 \times x_5 + x_5 \times x_1 \\ &+ x_1 \times x_3 + x_1 \times x_4 + x_3 \times x_5 + x_2 \times x_4 + x_2 \times x_5) \end{aligned} \tag{3.53}$$

In this case, we have five NN interactions and another five NNN interactions. Obviously, in this system all the spins are equivalent and there is frustration for many of the triplets of spins.

Figure 3.31 shows TI and CI for this case. Although there are numerous possibilities of frustration in this system, the values of CI at large and negative βJ is quite small.

The actual limiting values of TI and CI at $\beta J \to \infty$ depends on the limiting values of the SMI.

3.11 Six-Spin Systems

In this section, we examine only two cases of a six-spin system.

We start with a case of a perfect hexagon. Figure 3.32 shows the TI and CI for this case. Again, the values of TI and CI for $\beta J = 0$, and in the limits of $\pm\infty$, are as expected. Note that CI is everywhere *positive*.

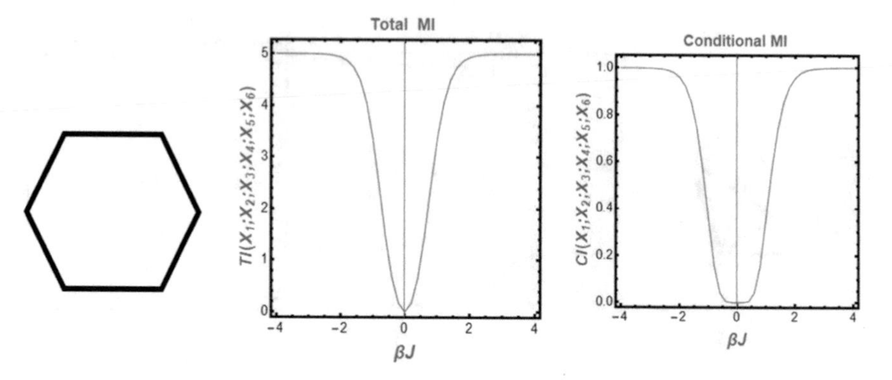

Fig. 3.32 The total and the conditional MI for the six-spins arranged in a regular hexagon, as a function of $\boldsymbol{\beta J}$

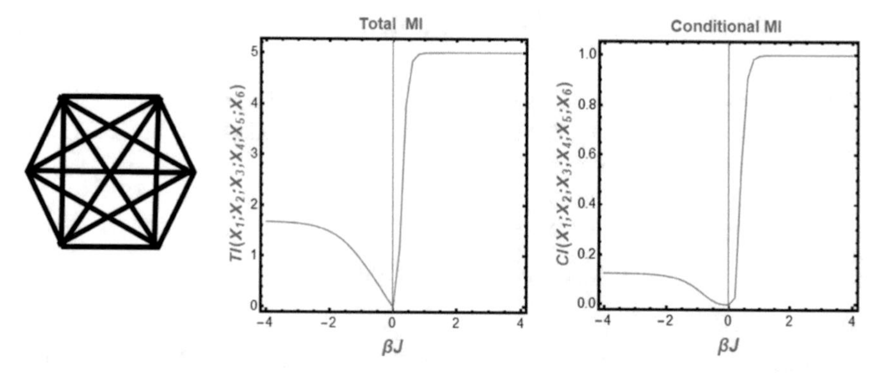

Fig. 3.33 The total and the conditional MI for the six-spins arranged in a regular hexagon with all pair interactions, as a function of $\boldsymbol{\beta J}$

Hexagon with All Pairs Interactions

Finally, we examine the case of six spins, each of which interacts with all the other spins, with the same parameter J. Figure 3.33 shows all the interactions between all the 15 pairs.

Figure 3.33 shows TI and CI. What is interesting is that the CI is everywhere *positive*. Although there are many triangles with frustrations, the CI does not show any negative values.

References

1. Ben-Naim, A. (2017). *Information theory, part I: An introduction to the fundamental concept*. World Scientific.
2. Fano, R. M. (1961). *Transmission of information; A statistical theory of communications*, MIT.

3. Matsuda, H. (2000). *Physical Review E, 62*, 3096.
4. Ben-Naim, A. (2015). *Discover probability. How to use it, how to avoid misusing it, and how it affects every aspect of your life*. World Scientific.
5. McGill, W. J. (1954). *Multivariate Information Transmission Psychometrika, 19*, 97.
6. Bell, J. E. (2003). *Evaluating Psychological Information*, 4th edition, Pearson.

Chapter 4
Entropy of Mixing and Entropy of Assimilation, an Informational Theoretical Approach

In Chap. 2 we discussed the entropy-changes that result from "turning on" the intermolecular interactions. In this chapter we discuss a few processes in ideal gases, where no change in the intermolecular interactions are involved. In these processes the changes in entropy is due either to changes in the accessible volume (e.g. mixing) or to changes in the number of indistinguishable particle (assimilation).

The process of mixing ideal gases features almost in any textbook on thermodynamics. It is *correctly* used to demonstrate the workings of the Second Law (entropy increases in a spontaneous process in an isolated system). It is *incorrectly* used as an example of the (wrong) interpretation of entropy as a measure of disorder. It is unfortunate that mixing is one of the most misunderstood and misinterpreted process in thermodynamics.

Sometime between 1986 and 1987, I submitted for publication an article on the "Entropy of Mixing." It was rejected by seven journals, until it was finally accepted and published in the American Journal of Physics, Ben-Naim [1, 2]. Interestingly, what enraged most of the reviewers of the article was not my claim that entropy *is not* a measure of disorder, but rather the more shocking claim that "Entropy of mixing" of ideal gas has nothing to do with the *mixing*! This was, and still is, a shocking revelation.

In discussing the Second Law of thermodynamics, most text book provide the example of the spontaneous process of mixing ideal gases, Fig. 4.1a. The general case of mixing process is the one where one starts with an isolated system having c-compartments. In compartment *i*, there are N_i molecules of type *i* (i.e. different molecules in different compartments). If we remove the partitions between the compartments, we *always* observe mixing. Once the molecules are mixed they will never de-mix spontaneously, or separate into regions each of which contains one component.

In numerous popular science books the authors tell you that this is an illustration of the Second Law of thermodynamics. They also add that that *mixing* is obviously a process of disordering, therefore this process proves that entropy increase is associated with an increase in disorder.

A. Ben-Naim, *Information Theory and Selected Applications*,
https://doi.org/10.1007/978-3-031-21276-5_4

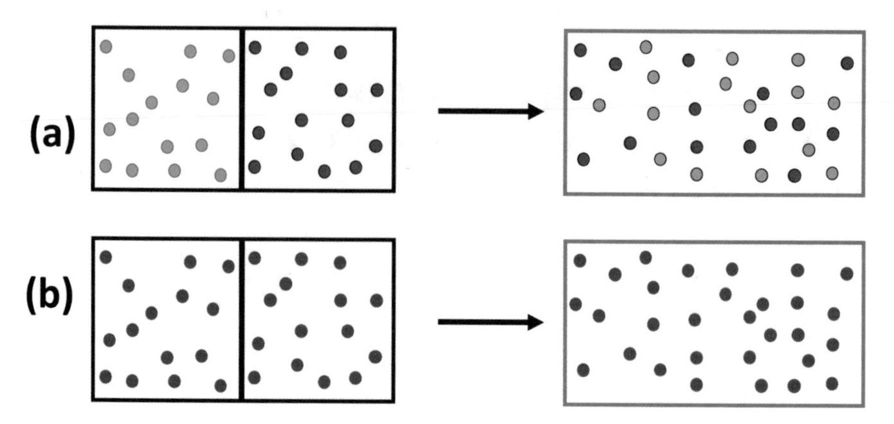

Fig. 4.1 Two processes of mixing ideal gases: **a** Mixing of different gases, **b** "mixing" of the same gases

Before we continue to discuss this process of mixing it should be emphasized that the statement: "mixing is always a spontaneous process" is, in general, not true. If we start with oil and water in two compartments, and remove the partition between the two phases, we shall not observe mixing. Therefore, from hereon we shall discuss only ideal gases. In this case, removal of the partition will always result in mixing of the gases, and in this particular process as shown in Fig. 4.1a, the entropy will always increase. In the following sections we shall discuss processes of mixing and de-mixing for which the entropy can either increase or decrease. We shall explore the entropy-change for some simple processes. In all of these we assume that the gases are ideal; no intermolecular interactions, hence, no change in energy (or temperature) in the process.

We start with the process of expansion of an ideal gas from volume V to $2V$. Then, we shall proceed to study mixing of two gases. This process was first studied by Gibbs [3], who erred in the interpretation of the "entropy of mixing" as well as in some of his conclusions. We remind the reader that entropy is a particular example of SMI. In general, SMI is not entropy (see Chap. 5 of Ben-Naim [4]). Therefore, all our interpretations of entropy increases will be based on viewing entropy as an SMI.

4.1 "Entropy of Mixing" of Two Different Ideal Gas

The reader should note that we enclosed the words "Entropy of Mixing" in quotation marks. The reason, as we shall soon see, is that the very term "entropy of mixing" is not appropriate for the change in entropy we observe in the processes shown in Fig. 4.1a.

Before we analyze and interpret this process we first discuss a simpler process; an expansion of ideal gas from V to $2\ V$, Fig. 4.2. It is straightforward to calculate

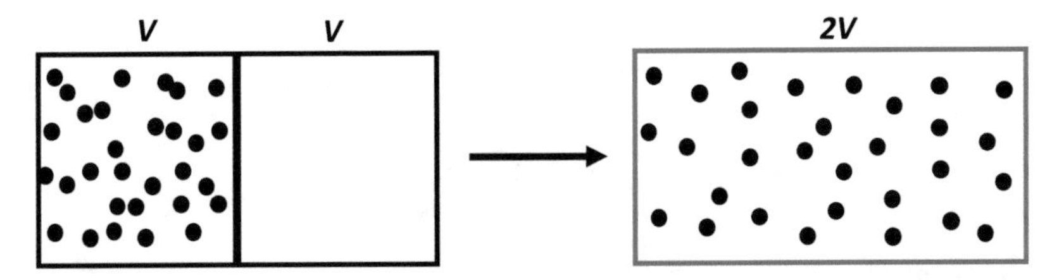

Fig. 4.2 A process of expansion of an ideal gas from V to 2 V

the entropy-change for this process. The partition function (PF) of an ideal gas of N particles at temperature T and volume V is:

$$Q(T, V, N) = \frac{q^N V^N}{\Lambda^{3N} N!} \tag{4.1}$$

Here, q is the internal function of each molecule, Λ^{3N} is the momentum PF. Both of these quantities will be unchanged in all the processes discussed in this chapter, for details see Ben-Naim [5, 6].

The change in the Helmholtz energy for the process of expansion in Fig. 4.2 is:

$$\Delta A = -k_B T \ln \frac{(2V)^N}{V^N} = -k_B T \ln 2^N \tag{4.2}$$

$$\Delta S = -\frac{\partial \Delta A}{\partial T} = k_B \ln 2^N = -k_B N \ln 2 \tag{4.3}$$

Dividing ΔS by $k_B \ln 2$, we obtain the change in the SMI for this process:

$$\Delta \text{SMI} = N \tag{4.4}$$

The interpretation of the entropy-change for the process of expansion in terms of SMI, is quite simple. Each particle was originally confined to a volume V. After the expansion, the accessible volume for each particle is 2 V. Therefore, in the process of expansion in Fig. 4.2 we lose one bit per particle, hence, N bits for N particles. We stress here that this result is *independent of the type* of particles (see next section for Gibbs's failure to understand this fact in the case of mixing of ideal gases). Clearly, this result is independent of the temperature. Unfortunately, some people who interpret entropy in terms of "spreading of energy" would say that the higher the temperature, the larger the spread of energy, and therefore, the larger the entropy-change should be! This is not a joke, it appears in some articles, and it was written by a reviewer of one of my articles. See Ben-Naim [7, 8].

Next, we proceed to the mixing process shown in Fig. 4.1a. We calculate the change in entropy for this process from the ratio of the PF in the final, and the initial final states in Fig. 4.1a, i.e.,

$$Q(initial\, state) = \frac{q_A^{N_A} q_B^{N_B} V^{N_A+N_B}}{N_A! N_B! \Lambda^{3N_A} \Lambda^{3N_B}}$$

$$Q(final\, state) = \frac{q_A^{N_A} q_B^{N_B} (2V)^{N_A+N_B}}{N_A! N_B! \Lambda^{3N_A} \Lambda^{3N_B}} \tag{4.5}$$

The change in the Helmholtz energy is:

$$\Delta A = -k_B T \ln \frac{Q(final)}{Q(initial)} = -k_B T (N_A + N_B) \ln 2 \tag{4.6}$$

For the specific process on Fig. 4.1a, we take one mole of A and one mole of B in each compartment; $N_A = N_B = N_V$, where $N_V = 6.02 \times 10^{23}$ is the Avogadro number. Hence, the entropy-change is:

$$\Delta S = -\frac{\partial \Delta A}{\partial T} = k_B 2N \ln 2 \tag{4.7}$$

What is the interpretation of this result?

Most authors of popular science books would explain:

1. Mixing is obviously a process in which disorder increases.
2. Entropy is a measure of disorder.

Therefore, they conclude, the positive change in this process is due to the *mixing*, i.e. increase in disorder. A generalization of this statement of the Second Law is: "*The disorder of the universe always increases.*"

Unfortunately, the above "argument," as well as the conclusion are faulty. We all accept, based on intuition that mixing is disorder. We do not have a definition of disorder. Therefore, such a statement can be said to be qualitatively correct. However, entropy is not a measure of disorder (for details and examples, see Ben-Naim [7, 8]). Therefore, the conclusion that increase in entropy is due to increase in disorder is faulty. In fact, as we shall soon see that the entropy-change, in the process as provided in Eq. (4.7), *is not* due to mixing. The last statement was shocking to many readers of my 1987 article.

Gibbs [3], who studied this process in details, expressed his puzzlement regarding the fact that the entropy-change in this process is independent of the type of molecules. Why should the entropy-change for mixing argon and neon be the same as mixing methane and ammonia, or water and alcohol gases?

Here is what Gibbs has to say about the entropy of mixing:

> But if such considerations explain why the mixture of gas-masses of the same kind, stands on different footing from mixtures of gas-masses of different kinds, the fact is not less significant than the increase of entropy due to mixture of gases of different kinds, in such a case as we have supposed, is independent of the nature of the gases.

Gibbs was puzzled by the fact that the "entropy of mixing" is independent of the "nature of the gases," more than the fact that the entropy-change becomes zero when the two gases in the two compartments are identical (see below for a discussion of this case). His bafflement was justified. When two gases react in a chemical reaction, even in the ideal phase, the resulting entropy-change will depend on the type of gases. When two liquids mix, the corresponding change in entropy is dependent on the type of molecules. Therefore, it is a real puzzle that the entropy-change in the mixing process in Fig. 4.1a is *independent* of the type of gases.

Gibbs believed that the entropy-change in the process of mixing, which he referred to as the "entropy of mixing" is due to the *mixing of the gases*. Unfortunately, he was wrong because he failed to realize that the "entropy of mixing" is due to the *expansion* of each of the gases from V to 2 V. As we have seen in Eq. (4.3), the entropy of expansion of one mole of a gas from V to 2 V is:

$$\Delta S(expansion) = k_B N \ln 2 \tag{4.8}$$

In the process of mixing, as can be seen from Eq. (4.6), the change in entropy is due to the *expansion* of one mole of A, from V to 2 V, and another *expansion* of one mole of B, from V to 2 V. Therefore, the change in entropy for this process is twice as large as in Eq. (4.8):

$$\Delta S(mixing in Fig.\ 4.1a) = k_B 2N \ln 2 \tag{4.9}$$

We can conclude that the process shown in Fig. 4.1a is equivalent to two processes of expansion as shown in Fig. 4.3. Clearly, what is called "entropy of mixing" is, in fact, entropy of *expansion*, and it has nothing to do with the *mixing process*. The interpretation of the entropy-change is the same as the interpretation of the entropy of expansion. In the process of mixing, Figs. 4.1a or 4.3a; each particle was initially confined to a volume V, and finally to the volume of 2 V. Therefore, for 2 N particles the change in the SMI is:

$$\Delta \text{SMI} = 2N \tag{4.10}$$

Compare this result with the Eq. (4.4). Had Gibbs understood that the entropy-change in the process of Fig. 4.1a is due only to the change in the *accessible volume* of each particle he would not have been puzzled by the fact that "the entropy of mixing" is independent of the "type of gases." It is now clear that the *mixing* of gases is not the cause of the so-called "entropy of mixing."

It is also appropriate to comment on the relevance of this process of mixing to the Second Law. Most of the textbooks would conclude that mixing, as a measure of disorder is an irreversible process, and hence would reformulate the Second Law as a tendency of the Universe to go from order to disorder. This statement is nonsense for several reasons: the entropy of the universe is not definable, and entropy is not a measure of disorder!

Most popular science books mention mixing of gases, or coffee and milk as example of the Second Law. This is unfortunate. One can easily construct mixing

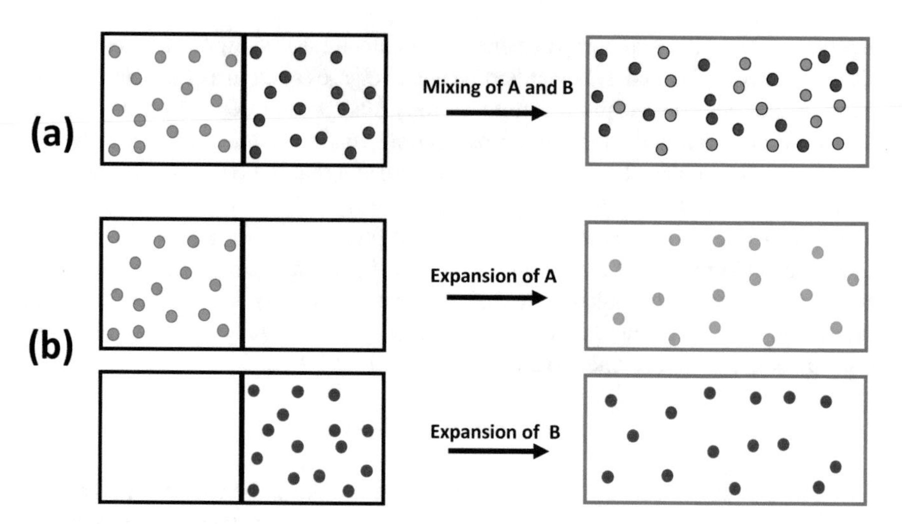

Fig. 4.3 The processes of mixing ideal gases (**a**) is equivalent to two processes of expansion (**b**)

processes of ideal gases with zero entropy-change (Fig. 4.4), and de-mixing of ideal gases with positive change of entropy (Fig. 4.5). Therefore, mixing by itself, is not a typical example of the Second Law. In Fig. 4.4 there is a mixing, but the change in the accessible volume per particle is zero. Therefore, the change in entropy is zero. In Fig. 4.5 we have de-mixing. Here, the volume accessible to each particle increases. Therefore, the entropy-change is positive (for more details on these processes, see Ben-Naim [7, 8]).

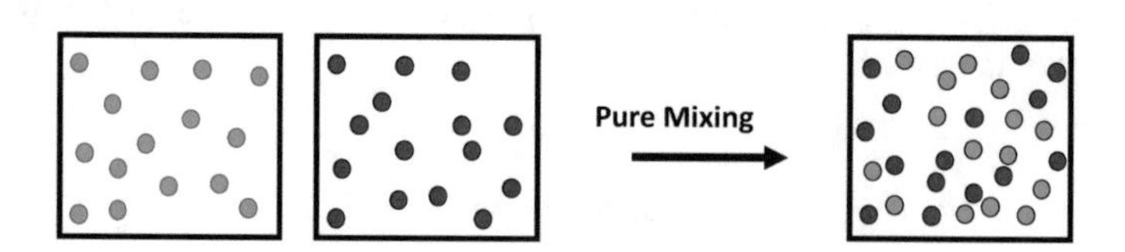

Fig. 4.4 A process of mixing with zero change in entropy

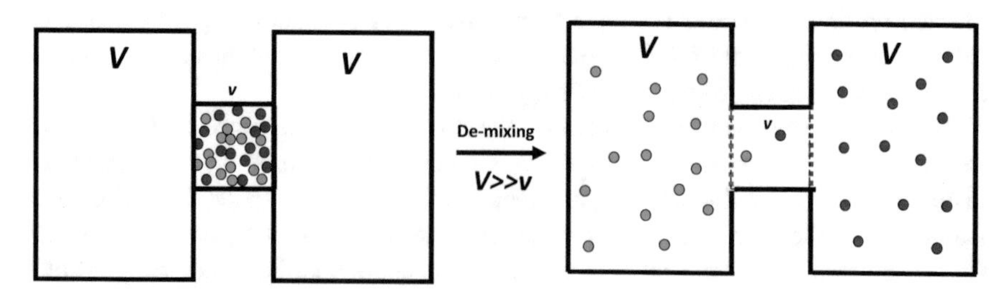

Fig. 4.5 A process of spontaneous de-mixing with positive change in entropy. The dotted red and blue partitions are permeable to red and blue particles respectively

In concluding this section, we emphasize that the "entropy of expansion" is a more appropriate term for the entropy-change in the process of Fig. 4.1a. The reason is that the entropy-change in this process is due only to the expansion of each gas, and has nothing to do with the mixing. Note that we discussed mixing of ideal gas. When there are intermolecular interactions the "entropy of mixing" depends on the "type of molecules," and it can be positive, negative or zero.

4.2 Entropy of Assimilation

Before we define the concept of assimilation, let us go back to Gibbs who analyzed the process of mixing. As we pointed out in Sect. 4.1, Gibbs was puzzled by the fact that the so-called "entropy of mixing" for the process shown in Fig. 4.1a is independent of the type of gases. Gibbs is also credited for the introduction of the factor $N!$ to account for the *indistinguishability* of the particles (see Ben-Naim [5–7]). Gibbs also understood why the entropy-change in the process of Fig. 4.1b is zero. Numerous authors who discuss the two processes in Fig. 4.1, noted that the entropy-change drops sharply and discontinuously from $k_B N \ln 2$ to zero when the two gases become *identical*. They referred to this puzzling fact as the Gibbs Paradox. In fact, there is no Gibbs Paradox, and Gibbs never saw any paradox in the fact that $\Delta S = 0$ in the process of Fig. 4.1b. Gibbs, as we noted in the previous section attributed the "entropy of mixing" in Fig. 4.1a to the *mixing*. In the process of Fig. 4.1b there is no *mixing*, hence we should not expect any "entropy of mixing." Here is what Gibbs had to say about this process.

> If we should bring into contact two masses of the same kind of gas, they would also mix but there would be no increase in entropy.
>
> When we say that when two different gases mix by diffusion ... and the entropy receives a certain increase, we mean that the gases could be separated and brought to the same volume...by means of certain changes in external bodies, for example, by the passage of a certain amount of heat from a warmer to a colder body. But when we say that when two gas masses of the same kind are mixed under similar circumstances, there is no change of energy or entropy, we do not mean that the gases which have been mixed can be separated without change to external bodies. On the contrary, the separation of the gases is entirely impossible.

Gibbs distinguished between the two processes in Fig. 4.1, by referring to the first as mixing of "two *different gases*," and to the second as mixing "two gas masses of the *same kind*." We shall use the term "mixing" (i.e. proper mixing) for the former, and "assimilation" for the latter, i.e. when two gases of the same kind are mixed. A more precise definition of assimilation will be discussed below.

In the quoted paragraph Gibbs states something quite interesting. The process in Fig. 4.1a, which we all know is an irreversible process, can be reversed. On the other hand, the process in Fig. 4.1b, which is a "non-process," or a "reversible process," cannot be reversed, "*on the contrary, the separation of the gases is entirely impossible.*"

Clearly, these statements sound paradoxical. How can an irreversible process be reversed, and a reversible process cannot be reversed? Gibbs understood that the first process may be reversed, but the reversal of the mixing in process 4.1a is not spontaneous. He also claimed that the reversal of the process in Fig. 4.1b is "*entirely impossible*." To understand why it is "*entirely impossible*," Gibbs introduced the idea that particles are indistinguishable (ID). In his view reversing process 4.1b requires bringing each particle originating from the left compartment, back to the left compartment, and each particle originating from the right compartment, back to the right compartment. Such a process is impossible because the particles are ID. He probably tried to follow, mentally each particle moving in the system after removing the partition in process Fig. 4.1b. Clearly, because of the ID of the particle there is no way of reversing process 4.1b; there is no way we can know from which compartment each specific particle originated. Hence, such a process is "*entirely impossible*."

Gibbs argument was quite convincing. Indeed, particles are ID. It is also clear that it is not possible to know from which compartment each particle originated. However, Gibbs erred in concluding that the *reversal of the process* in Fig. 4.1b is "*entirely impossible*." Why? Ironically, the answer is based the same argument used by Gibbs to conclude that the reversal of the Process 4.1b is "*entirely impossible*," i.e. because the particles are ID. If the particles are ID, it follows that the reversal of this process is "*trivially possible*." How? Simply returning the partition to its original place, and we shall have reversed the process, see Fig. 4.6. Thus, although Gibbs understood that particles are ID, he failed to understand the implication of this indistinguishability. Putting back the partition brings the system to a state which is indistinguishable from the initial state in process 4.1b.

Most textbooks on thermodynamics discuss the two processes in Fig. 4.1 and concluded that:

1. In 4.1a, we *see* mixing, therefore the so-called "entropy of mixing" is the change in entropy due to *mixing*.
2. In 4.1b, we do not *see* any process occurring upon the removal of the partition. No process, hence no change in entropy! Simple and straightforward conclusions, but wrong.

As we have seen in the previous section the entropy-change in Process 4.1a is due to expansion, not to the mixing. Second, it is not true that Process 4.1b is a "non-process." In fact, it is a combination of two processes. To see that we write the PF of the system before and after the process in Fig. 4.1b:

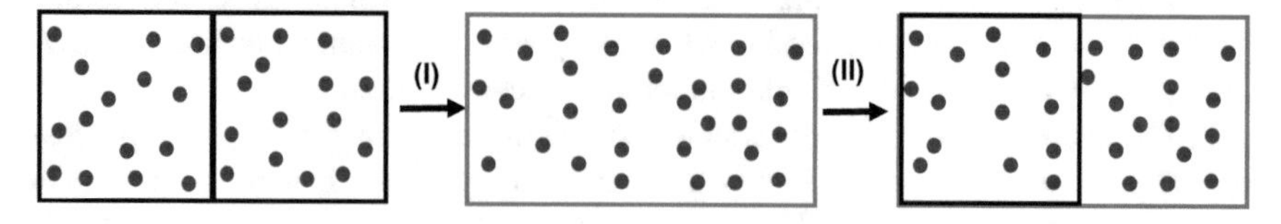

Fig. 4.6 To reverse the process in Fig. 4.1b, (I), simply place the partition back in its place, (II)

$$Q(initial\,state) = \frac{q^N q^N V^N V^N}{\Lambda^{3N} \Lambda^{3N} N!N!} \tag{4.11}$$

$$Q(final\,state) = \frac{q^{2N} (2V)^{2N}}{\Lambda^{6N} (2N)!} \tag{4.12}$$

The Helmholtz energy change is:

$$\begin{aligned} \Delta A &= -k_B T \ln \frac{Q(finalstate)}{Q(initialstate)} \\ &= -k_B T \ln \left[\frac{(2V)^{2N} (N!)^2}{(V)^{2N} (2N)!} \right] \end{aligned} \tag{4.13}$$

The entropy-change is:

$$\Delta S = -\frac{\partial \Delta S}{\partial T} = k_B 2N \ln \frac{2V}{V} - k_B \ln \frac{(2N)!}{(N!)^2} \tag{4.14}$$

Note that this change in entropy is always *positive*, not zero as expected from a "non-process." This follows from the identity:

$$2^{2N} = (1+1)^{2N} = \sum_{i=1}^{2N} \binom{2N}{i} > \frac{(2N)!}{N!N!} \tag{4.15}$$

Note that the sum in (4.15) is over 2 N positive terms. This sum is therefore larger than one of its terms: $\binom{2N}{N}$.

Rearranging (4.15) we obtain:

$$2N \ln 2 - \ln \frac{(2N)!}{(N!)^2} > 0 \tag{4.16}$$

Therefore, the entropy-change in (4.14), as well as the change in the SMI in process 4.1b is always positive.

In thermodynamics we deal with systems with very large number of particles. Thus, using the Stirling approximation for $\ln N!$ and $\ln(2N)!$ we find that in the limit of large *N:*

$$\begin{aligned} \lim_{N \to \infty} \Delta S &= 2N \ln \frac{2V}{V} - [2N \ln 2N - 2N \ln N] \\ &= 2N \ln 2 - 2N \ln 2 = 0 \end{aligned} \tag{4.17}$$

This means that for any *finite* number of particles the change in the SMI (i.e. the ΔS in Eq. (4.14) divided by $k_B \ln 2$) is always positive. It is only for very large number of particles that the two terms on the right-hand-side of Eq. (4.14) cancel each other.

Examining the two terms on the right-hand-side of Eq. (4.14) we realize that the first term is exactly the same as the entropy-change in the mixing process of Fig. 4.1a. Indeed, in the process 4.1b each particle "expands" its accessible volume from V into 2 V. However, in contrast to the process 4.1a where we had *only expansion*, in the process of Fig. 4.1b we have another term due to the change in the number of indistinguishable particles (initially, we had two compartments each with N, ID particles, and in the final state we have 2 N, ID particles). We refer to the second term on the right-hand-side of (4.14) as the *assimilation* entropy. Thus, the two contributions to the entropy-change in (4.14) are due to two processes; expansion and assimilation. These two terms cancel each other only for very large N.

In concluding this session, we can say that in Fig. 4.1b, we *see* a non-process, but in fact, it consists of two processes; expansion and assimilation. Unlike the case of mixing in which there is only one contribution to ΔS—the expansion, here we have two contributions: one is due to expansion which is positive, and is exactly the same as the contribution to the entropy-change in process 4.1a. The second is negative, which is due to assimilation. The concept of assimilation, which is more fundamental than mixing, has unfortunately not yet been assimilated in the minds of scientists. It has been more than 30 years since the publication of my first article on this topic, and yet people still write that the "entropy of mixing" is due to the mixing process, and that mixing is essentially an irreversible process.

4.3 Is There a Pure De-assimilation Process?

In Figs. 4.4 and 4.5 we saw that there exists a process of mixing in which there is no change in entropy, and there exists a reversal of the mixing process which we call de-mixing with a positive change in entropy. The process in Fig. 4.4 is often called "reversible" mixing. I prefer to call it "pure mixing" process. It is *pure* in the sense that only *mixing* occurs in this process; no change in either the locational or the momentum distributions. The two questions we now pose are: Is there a "pure" assimilation process? Is there a spontaneous reversed-process of assimilation?

A possible "*pure assimilation*" process is shown in Fig. 4.7. Here, there is no change in either the locational or the momentum distribution. This process involve N, ID particles, and another N, ID particles, assimilating into 2 N, ID particles. We already know that this process is associated with negative change in entropy (see second term on the right-hand-side of Eq. 4.14). Therefore, we can expect that such a process will not occur spontaneously.

Next, we discuss the reverse-assimilation process. Formally, if we reverse the process shown in Fig. 4.7a, we obtain the process in Fig. 4.7b. This process is associated with positive change in entropy. We shall refer to this process as "*pure de-assimilation*," i.e. a system of 2 N particles splits into two subsystems, each having

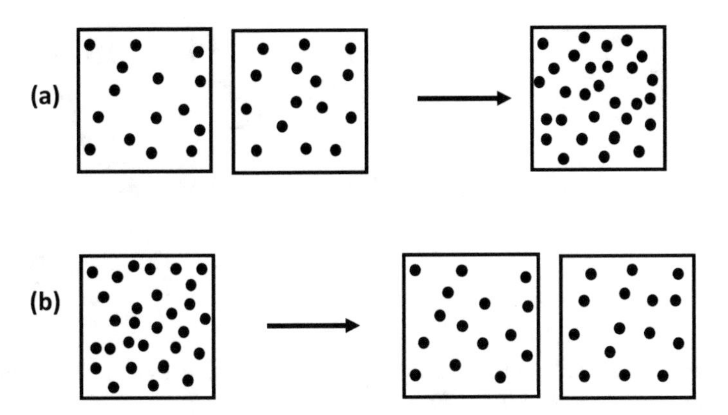

Fig. 4.7 **a** A pure assimilation process and **b** A pure de-assimilation process of an ideal gas

the same volume and energy, and contains N particles. Obviously, such process never occurs spontaneously. The Second Law of thermodynamics states that when a spontaneous process occurs in an isolated system the entropy will increase. It does not state that any process for which we calculate a positive change of entropy must occur spontaneously. However, we can still ask: Can we design a pure de-assimilation process which occurs spontaneously? The answer is Yes!

The process shown in Fig. 4.7b is a pure de-assimilation process, but it does not occur spontaneously. We now show an equivalent process which we can observe occurring spontaneously.

4.4 Racemization as a Pure De-assimilation Process

By *pure de-assimilation* process, we mean a process for which the accessible volume for each particle does not change. The velocity distribution (or the temperature) does not change, and the energy of the system does not change.

We start by considering a chemical reaction of isomerization. Figure 4.8 shows two configurations of a molecule having the same chemical formula, say, 1, 2-dichloroethene.

We start with one isomer, say the Cis one on the left hand side of Fig. 4.8 and add a catalyst. The system will approach a new equilibrium state in which we shall have an equilibrium mixture of Cis and Trans. The entropy-change in this process has two contributions; one, due to changes in the internal energies of the molecules, and the second, due to the change in the numbers of the Cis and Trans isomers.

There is one very special case of an *isomerization* reaction for which no changes in the internal energies of the molecules occur. This is the case of the two isomers which are mirror images of each other. These are referred to as two enantiomers. Figure 4.9 shows an example of such a molecule; *d* and *l* alanine. A carbon atom has

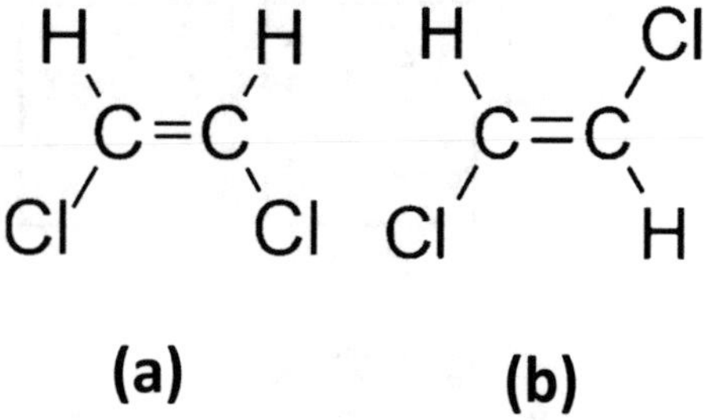

Fig. 4.8 Two isomers of 1,2-di-chloroethene: **a** cis and **b** trans

four different neighboring groups. Such a molecule is said to have a chiral center. As we can see from the Fig. 4.9, the two isomers are identical, except for being mirror images of each other.

The history of such molecules is truly fascinating. See Ben-Naim [9] for details. These two compounds have exactly the same properties except for the rotating plane of polarized light in different directions. The two isomers are denoted *levo* (l) and *dextro* (d), for rotating polarized light to the left and right, respectively.

Suppose we start with a pure l form, and add a catalyst, which enables the conversion between the two isomers. The system will evolve into a mixture of d and l, Fig. 4.10. At equilibrium we will find equal numbers of l and d molecules. Since the two isomers have the same internal energies, the entropy-change for this reaction is due to only changes in the number of molecules of d and l forms (here, from initially 2 N, of the l form to finally, N, l and N, d forms). The entropy-change for this reaction is $2Nk_B \ln 2$. Here are three possible interpretations of this entropy-change.

(i) *The thermodynamic explanation* is, according to many text books, that this entropy-change is due to "mixing." Some authors refer to this entropy-change

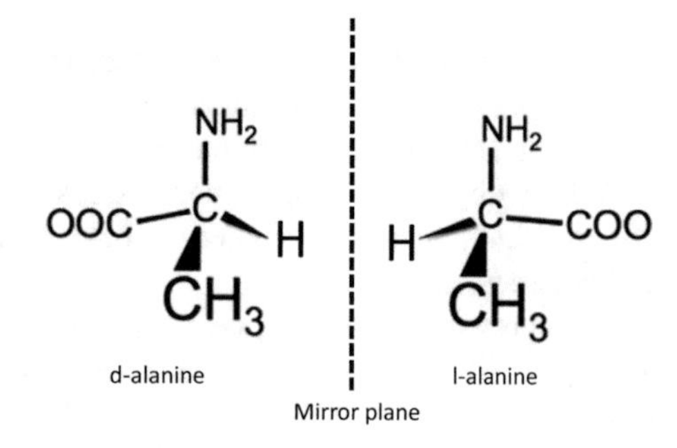

Fig. 4.9 Two enantiomers of alanine are mirror images of each other

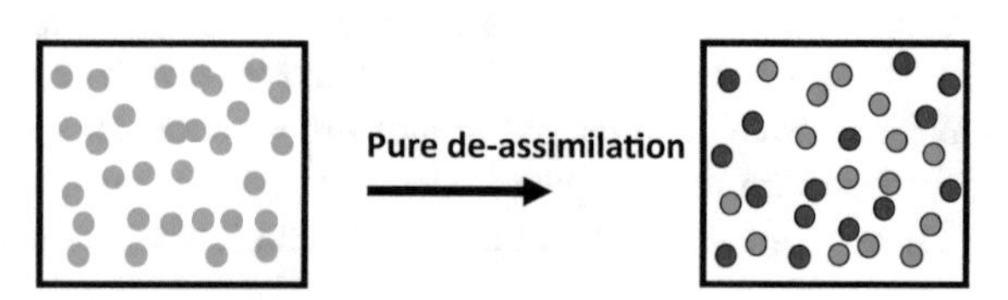

Fig. 4.10 The process of a spontaneous pure de-assimilation

as "entropy of mixing." Indeed, the quantity $2Nk_B \ln 2$ is exactly the same as the entropy-change in the process shown in Fig. 4.1a. As we have explained the term "entropy of mixing" is not appropriate for the process in Fig. 4.1a. Although we *observe* a mixing of two components, the entropy-change in this process is *not* due to the mixing but to the expansion of the two gases from V to 2 V.

The term "entropy of mixing" is *a fortiori* inappropriate for the process of racemization, Fig. 4.10. Here, we do not *observe* a mixing of two components, l and d, but an *evolution* of a pure l into a mixture of two components. Thus, thermodynamics cannot explain the origin of the racemization process.

Thermodynamics can only offer a qualitative explanation for the equal amounts of d and l at equilibrium. The equilibrium constant in this case is related to the standard Helmholtz energy for this reaction. Since the two isomers have the same internal energy levels, the standard Helmholtz energy must be zero. Hence, the equilibrium constant must be one, hence N_l and N_d must be equal at equilibrium.

(ii) *The statistical mechanical explanation* is simple. For an ideal gas we can calculate the partition function in the initial and the final states, then calculate the entropy-change. We find that the entropy-change $2Nk_B \ln 2$ is due only to changes in the numbers of l and d forms. As for the equilibrium ratio of N_l and N_d, statistical mechanics provides us with the equation for the equilibrium constant:

$$K = \left(\frac{N_l}{N_d}\right)_{eq} = \exp\frac{q_l}{q_d} = 1$$

where q_l and q_d are the internal partition functions of l and d isomers, respectively. Since the internal energies of the two isomers are the same, we must have $q_l = q_d$ and $K = 1$.

(iii) *The informational interpretation* is the same as statistical interpretation, i.e. the change in entropy is due to changes in the *identity* of N molecules from l to d. We referred to this process as pure de-assimilation process for the following reasons: First, the accessible volume for each particle does not change in the process (it is V initially, and finally). Second, there are no energetic changes, and neither the temperature, nor the velocity distribution change in the process. The only change that occurred is that half of the molecules *acquired a new identity*. We started with 2 N, ID particles and ended with two groups of N ID particles. This is therefore equivalent to the process shown in Fig. 4.7b.

Recall that in the process of expansion from V to 2 V we claimed that the change in entropy is due to the loss of one bit per particle. Similarly, in the process of mixing (of ideal gases, Fig. 4.1a) there is a loss of one bit per particle. The mixing does not contribute to the entropy-change. Exactly the same explanation applies to the entropy-change in the process of de-assimilation (either in Fig. 4.7a or in 4.10); we have a loss of one bit per particle. For instance, in the process of racemization, in the process there is no change in the volume accessible to each particle nor change in the velocity distribution. However, if you pick up a particle at random then, initially you knew its identity (*l* in Fig. 4.10), but after the process the particle may be either *l* or *d*. hence there is a loss of one bit per particle in this process.

4.5 An Example of the Entropy Formulation of the Second Law

Consider a more general process of mixing. We start with an isolated system with *c*—compartments. Initially, in compartment *i*, there are N_i particles in volume V_i. The total volume and the number of particles are:

$$V = \sum_{i=1}^{c} V_i \tag{4.18}$$

$$N = \sum_{i=1}^{c} N_i \tag{4.19}$$

From Shannon's theorem (see Sect. 2.11.1 of Ben-Naim [4]) we know that the distribution that maximizes the SMI is the uniform distribution. We also know that the uniform distribution of locations is the equilibrium distribution of any ideal gas system. Thus, we know that initially the locational distribution in each compartment is uniform and equal to $1/V_i$. We also know that the equilibrium locational distribution after removing all the partitions will be 1/*V*, i.e. uniform probability of finding a particle in any location in the entire volume *V*.

We now show that when we remove all the partitions in the system, the entropy must increase (note that there are no interactions among the particles and the process is carried out at constant *N*, *V*, *E* which means an isolated system).

We start with the Sackur-Tetrode equation for the entropy of ideal gas (see Sect. 5.2 of Ben-Naim [4]).

$$S(T, V, N) = k_B N \ln\left(\frac{V}{N}\alpha\right) + \frac{5}{2}k_B N \tag{4.20}$$

In Eq. 4.20 we collected all terms which do not change in the process into the constant α. Since both the initial and the final states in the process are equilibrium states, we can write the entropy of the initial and the final states as:

$$S(initial) = \sum_{i=1}^{c} k_B N_i \ln\left(\frac{V_i}{N_i}\alpha\right) + \frac{5}{2}k_B N_i \tag{4.21}$$

$$S(final) = k_B N \ln\left(\frac{V}{N}\alpha\right) + \frac{5}{2}k_B N \tag{4.22}$$

We now define the mole fraction and the volume fraction by:

$$y_i = \frac{V_i}{V}, \ x_i = \frac{N_i}{N} \tag{4.23}$$

With these two fractions we can write the entropy-change in the process as:

$$\Delta S = S(final) - S(initial)$$

$$= k_B N \sum_{i=1}^{c} x_i \ln \frac{x_i}{y_i} \geq 0 \tag{4.24}$$

By dividing ΔS by the constant $k_B \ln 2$, we obtain the Kullback–Leibler inequality [see Sect. 3.3 of Ben-Naim [4]). Thus, we have seen that starting from any distribution of particles (N_i in compartment i) and by removing all the partitions the entropy will always increase. This is a special case of the entropy-formulation of the Second Law. The proof was based on Shannon's theorem about the locational distribution of ideal gas at equilibrium.

In practice, we usually work with systems which are either at constant temperature and volume, or temperature and pressure. In the next section we present an example of the Gibbs energy formulation of the Second Law.

4.6 An Example of Gibbs Energy-Formulation of the Second Law

For a system at constant temperature, pressure, and total number of particles we have the following formulation of the Second Law:

> The Gibbs energy of an equilibrium system is minimum over all possible Gibbs energies of constrained equilibrium states of the same system having the same temperature, pressure, and total number of particles.

In the previous section we discussed a spontaneous process of mixing in an isolated system. Here we discuss a spontaneous process in a constant temperature and pressure. If we start with a constrained equilibrium system and remove the constraints keeping T, P, N constant the Gibbs energy of the system must decrease.

We discuss a special process which is of importance in the theory of protein-folding.

Suppose we have a molecule which can be in c-states (the analogue of c-compartments). Suppose also that in the presence of an inhibitor (the analogue of the partitions in Fig. 4.1) we can prepare a constrained equilibrium system with N_i

molecules in state i. An example would be a protein having a discrete number of conformation.

If we remove the inhibitor (the analogue of removal of the partition in the mixing process of Fig. 4.1) the system will change to a new equilibrium state. If we keep the temperature, the pressure and the total number of molecules fixed then the Gibbs energy of the system will *decrease* in this system. Specifically, suppose that we initially prepare the system with N_i molecules in state i, such that $\sum N_i = N$. If there is an inhibitor that inhibits the conversion between states, the system will stay in this constrained equilibrium state. The distribution of molecules in the different states can be represented by the vector $(x_1, x_2, \ldots, x_c)$ where $x_1 = N_i/N$ is the mole fraction of molecules in the state i.

If we remove the inhibitor, the system will attain a new equilibrium state. One can prove that there exists a single equilibrium vector $\left(x_1^{eq}, x_2^{eq}, \ldots, x_c^{eq}\right)$ such that the Gibbs energy change in the process will be:

$$\Delta G = k_B T N \sum_{i=1}^{c} x_i \ln \frac{x_i^{eq}}{x_i} \leq 0 \tag{4.25}$$

This is again the Kullback–Leibler inequality for the "distance" between the initial and the final vectors $(x_1, \ldots, x_c)$ and $\left(x_1^{eq}, \ldots, x_c^{eq}\right)$, respectively.

We have discussed here the Gibbs formulation of the Second Law for a discrete distribution of particles in different states. One can also go to the limit of continuous distribution of molecules in different states. The result is the same as the result shown in Eq. (4.25).

Details of the proof (4.25) and its relevance to the theory of protein folding are discussed in Ben-Naim [10].

4.7 A Baffling Experiments in Systems of Interacting Particles

In this section we briefly mention two baffling experiment. For more details, the reader is referred to Ben-Naim [11]. In this Chapter we studied two apparently very different processes. One is an expansion of an ideal gas from V to 2 V. The second is mixing of two different gases. In both of this processes we saw that the main "driving force" is the tendency of the locational distribution function to evolve into a distribution that maximizes the SMI. In both processes, at the instant we remove the partition we have a distribution which is a step-function with a constant value in one compartment, and zero value in the second compartment. This distribution starts to evolve right after the removal of the partition until we reach a final distribution which is uniform in the entire new volume of 2 V. This final distribution is also the equilibrium one. This is the reason why we identify the distribution which maximizes the SMI with the equilibrium distribution. This is also why we identify the tendency

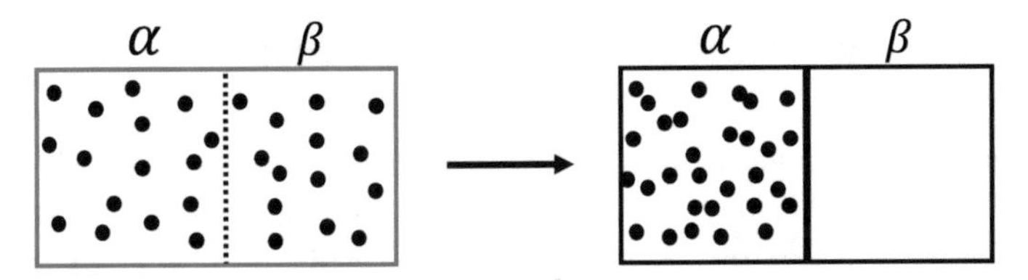

Fig. 4.11 An apparent "reversal" of the process of expansion shown in Fig. 4.2

to evolve into the new equilibrium state as one manifestation of the Second Law of thermodynamics.

We say that these two processes are irreversible—meaning that we never observe the reversal of these processes. A gas occupying the volume 2 V will never be found in a smaller region, say, in one compartment of volume V, and a mixture of two gases will never separate into two places each containing only one component.

All that has been said above regarding ideal gases is true. It is, in general not true when there are strong intermolecular interactions. Here, we describe in a qualitative manner how a seemingly reversal of both of the processes of expansion and mixing can occur. A more quantitative discussion of such processes is available in Ben-Naim [11].

Supposed we start with a system of two compartments, Fig. 4.11. The left contains N_A molecules of type A in a solvent α. The right contains N_A molecules of type A in a solvent β. The two solvents are transparent, but A is blue.

Initially, all the 2 N_A molecules are distributed in the entire system of volume 2 V. We now remove the constraint that allows the passage of A molecules between the two compartments. If A interacts very strongly and attractively with the solvent α, then we shall observe that most of the A molecules will move into the left compartment. The "driving force" for this process is the strong solvation of A in α. Since the two solvents are transparent, an observer who is not aware of the existence of the solvents α and β will "see" that the blue molecules A move from occupying the entire volume 2 V into the smaller region of volume V. This will seem to the observer as a "reversal" of the expansion process.

Similarly, one can start with mixture of A and B in the two compartments as on the right hand side of Fig. 4.1a. We take a solvent α which interacts favorably with A molecules, and a solvent β which interacts favorably with B molecules. Again, we assume that the two solvents α and β are transparent. The A-molecules are blue and the B-molecules are yellow. Therefore, initially we shall see a uniform green color in the entire system.

Once we remove the constraint allowing both A and B to move between the two compartments, we shall observe that the left compartment becomes blue and the right compartment becomes yellow. To an observer who is not aware of the existence of the two solvents, this process will seem to be a "reversal" of the mixing process.

The moral of these experiments is that one should not trust one's eyes when studying experiments of the type shown in Fig. 4.1. What you *see* is not always the cause of the process you observe.

4.8 Communal SMI and Communal Entropy

The concept of communal entropy has featured within the lattice models of liquids and mixtures. In this section we show first that this communal SMI is due to a combination of assimilation and expansion. And second, that this communal SMI turns to entropy only when the number of particles in each cell is very large.

Figure 4.12 depicts a process which called *delocalization.* Initially, we have N particles, each confined to a cell of size v. We remove all the partitions and the particles are allowed to occupy the entire volume V. The change in entropy in this process is:

$$\begin{aligned}\Delta S &= S(ideal\ gas) - S(localized)\\ &= kN\ln\left(\frac{V}{\Lambda^3}\right) - k\ln N! + \frac{3kN}{2}\\ &- kN\ln(v/\Lambda^3) - \frac{3kN}{2}\end{aligned} \tag{4.26}$$

Here $v = V/N$ is the volume of each call. Application of the Stirling approximation to ln $N!$ yields:

$$\Delta S = kN \tag{4.27}$$

The quantity kN, in Eq. (4.27) is known as the *communal entropy,* or the "delocalization" entropy. In fact, ΔS in Eq. (4.27) is the sum of two effects; the increase of the *accessible volume,* for each particle from v to V, and the *assimilation* of N particles. These two contributions are explicitly given by:

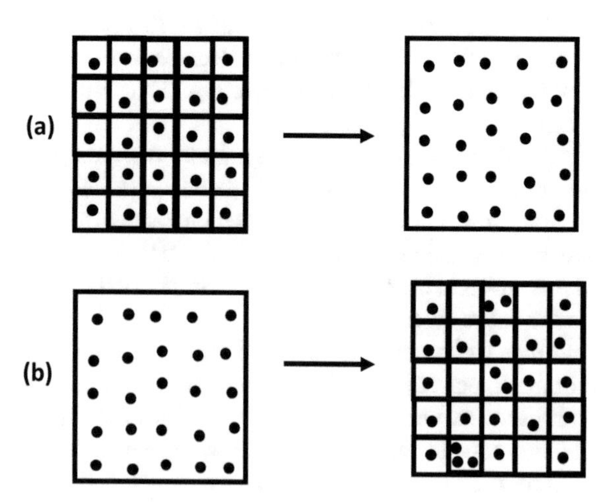

Fig. 4.12 **a** Delocalization process and **b** the "reverse" of process (**a**)

$$\Delta S = k \ln\left(\frac{V}{v}\right) - k \ln N! \tag{4.28}$$

Note that the volume per particle is denoted by $v = V/N$. Clearly, first term on the right hand side of (4.28) is the dominating one, i.e., the contribution of the volume change accessible to each particle is larger than the assimilation contribution.

Using the Stirling approximation, we get approximately:

$$\Delta S = kN \ln N - (kN \ln N - kN) = kN \tag{4.29}$$

Originally, the concept of "*communal entropy*" was introduced to explain the entropy-change in the process of melting, i.e. going from order to disorder. Specifically, it was stated that "*this communal sharing of volume gives rise to an entropy of fusion.*" Later, this idea was criticized by Rice [12]. Nowadays, the whole concept of "communal entropy" is considered to be obsolete.

Thus, the "communal entropy" is not a result of *volume change* only, but a combination of volume change and *assimilation.* This quantity, up to a multiplicative constant, is the change in the locational SMI. Note that in writing of Eq. (4.26) we assume that we can assign an entropy value for each cell containing one particle only. This is not completely justified as we will see below.

We now extend this experiment as follows: Instead of the initial state shown in Fig. 4.12, where there is only one particle in each cell, we assume that initially we have n particles in each cell. Thus, the total number of particles is nM, where M is the number of cells.

The change in SMI for the general process is:

$$\Delta \text{SMI} = nM \ln M + \ln \frac{(n!)^M}{(nM)!} \tag{4.30}$$

where we use the natural logarithm and $k = 1$.

In Eq. (4.30), there are two contributions to the SMI. The first due to the "expansion," i.e. each particle expands its accessible volume from V to $Mv = V$. The second term is due to assimilation. Initially, we have M cells, in each n indistinguishable particles. In the final state we have nM indistinguishable particles in the entire volume V.

We now examine the two extreme cases:

1. For $n = 1$, we have:

$$\Delta \text{SMI} = M \ln M - \ln M! \tag{4.31}$$

   Assuming that $N = M$ is large enough so that we can use the Stirling approximation as $\ln M! = M \ln M - M$, we obtain:

$$\Delta \text{SMI} = M \tag{4.32}$$

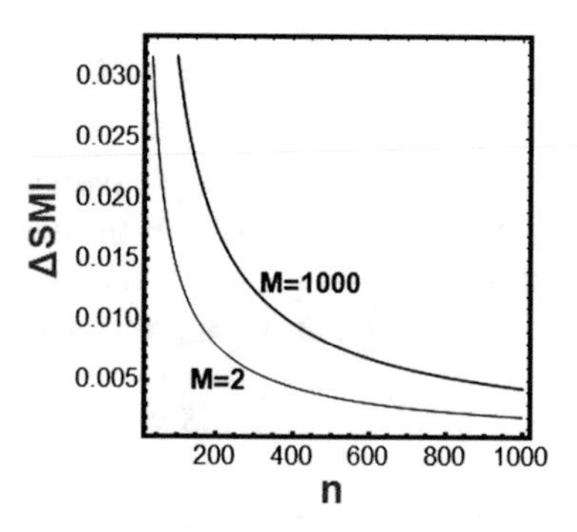

Fig. 4.13 The exact values of ΔSMI as a function of n, for two values of M

This is exactly the case we discussed above where we had $N = M$.

2. For larger n, such that we can use the Stirling approximation for both n and M, we have:

$$\Delta\text{SMI} \approx nM \ln M + M(n \ln n - n) - (nM \ln(nM) - nM) = 0 \quad (4.33)$$

In this case, we have $\Delta\text{SMI} = 0$, i.e. the change in SMI due to expansion is cancelled by the change due to the assimilation. Thus, we see that there is a fundamental difference between the case $n = 1$ (or any small number of particles), and large n. In both cases, we can define the entropy corresponding to the SMI by simply multiplying by a k_B (or also changing to $\log_2$).

Figure 4.13 shows the exact values of ΔSMI as a function of n, for a fixed value of M. We see that for any M the values of ΔSMI tend to zero at large n. We can now reverse the process in Fig. 4.12a and reach an important conclusion regarding a fundamental assumption used in the theory of irreversible thermodynamics.

Suppose we start with nM particles in a volume V, and put back the partitions to form the right-hand-side of Fig. 4.12b. Of course, we shall not get the same as the initial state. But to a good approximation we can say that on *average* there will be about n particles per cell. Now, if M is fixed and n is very large, then by putting back all the partitions, we have $S(initial) = MS(cell)$, i.e. we have the additivity of the entropy. However, when n is of the order of one, we will not have:

$$S(initial) \neq MS(cell)$$

Thus, for small n we lose the additivity of the entropy. One of the main assumptions of irreversible thermodynamics is that the entropy density $s(u, v, n)$ has the same dependence on the variables:u, v, n as the entropy of a macroscopic system: $S(U, V, N)$. Here, we saw that this is not so when n is small.

References

1. Ben-Naim, A. (1987). Is mixing a thermodynamic process? *American Journal of Physics, 55*, 725.
2. Ben-Naim, A. (1987). Mixing and assimilation in systems of interacting particles. *American Journal of Physics, 55*, 105–109.
3. Gibbs, J. W. (1906). *Collected scientific papers of J. Willard Gibbs*. Longmans.
4. Ben-Naim, A. (2017). *Information theory, part I: An introduction to the fundamental concept*. World Scientific.
5. Ben-Naim, A. (1992). *Statistical thermodynamics for chemists and biochemists*. Plenum Press.
6. Ben-Naim, A., & Casadei, D. (2016). *Modern thermodynamics*. World Scientific.
7. Ben-Naim, A. (2012). *Entropy and the second law. Interpretation and miss-interpretations*. World Scientific.
8. Ben-Naim, A. (2020). *Entropy: The greatest blunder ever in the history of science*. Independent Publisher, Amazon.
9. Ben-Naim, A. (2016). *Entropy, the truth, the whole truth and nothing but the truth*. World Scientific.
10. Ben-Naim, A. (2016). *Myths and verities in protein folding theories*. World Scientific.
11. Ben-Naim, A. (2008). *A farewell to entropy: Statistical thermodynamics based on information*. World Scientific.
12. Rice, O. K. (1938). *The Journal of Chemical Physics, 9*, 1.

Chapter 5
Information Transmission Between Molecules in Binding Systems

This chapter deals with relatively simple systems; macromolecules, usually proteins, each of which has a number of binding sites. Each site can bind one ligand molecule. The description of the system and the method of obtaining the various probabilities will be discussed in Sect. 5.1. Here, we discuss the motivation for writing this chapter as an example of application of IT.

It is well known that proteins are molecular machines that achieve some outstanding functions. Examples are the efficient transport of oxygen by hemoglobin and the regulation of biochemical processes by allosteric enzymes. The structures as well as the functions of these molecules have been studied in great detail. It was long recognized that the key factor which enables these molecules to achieve such an outstanding performance, is the possibility of *communicating information* between ligands occupying different sites.

Earlier studies of these systems focused on the so-called *cooperativity* of the binding process Ben-Naim [1]. Here, we use a more general terminology from IT to study the same systems.

Before discussing some specific models, we present here two examples of binding systems, with the help of which we can understand, in a qualitative manner, how the communication between ligands is responsible for the outstanding performance of these molecules.

The first example is the hemoglobin molecule. This molecule is perhaps the most studied protein. It was known for a long time that it is an efficient transporter of oxygen from the lungs to the cells in which oxygen is consumed. What do we mean by "efficient" transporter of oxygen, and how is this efficiency achieved?

Suppose you need to transport oxygen between two points, let us call them loading and the unloading terminals. The carrier, here the hemoglobin molecules (H_b) loads oxygen at the loading terminal and unloads it at the unloading terminal. The amount of oxygen being loaded on each H_b molecule depends on the pressure of the oxygen at the loading point, which is usually in the lungs. The larger the pressure the more oxygen will be loaded on the H_b. This relationship between the partial pressure of oxygen P_{o_2}, and the fraction of binding sites occupied (θ) is called the binding

A. Ben-Naim, *Information Theory and Selected Applications*,
https://doi.org/10.1007/978-3-031-21276-5_5

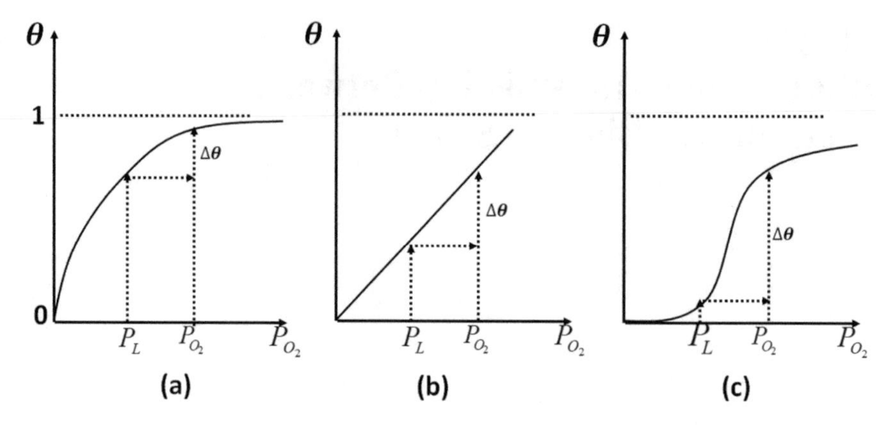

Fig. 5.1 Three possible forms of binding isotherms **a** Langmuir curve, **b** linear, and **c** cooperative curve (similar to the binding of oxygen to hemoglobin)

curve, or the binding isotherm (we assume that the temperature is the same at both terminals). Some typical binding isotherms are shown in Fig. 5.1. In all these curves we observe that the larger the pressure the larger the value of θ. The fraction θ is simply the average number of oxygen molecules, divided by the number of binding sites.

Once the H_b is loaded in the lungs it will be carried out by the blood to the cells, in which it will be unloaded and consumed. The unloading of the oxygen in the cell also depends on the local pressure. The lower the pressure of oxygen the larger amount of oxygen will be unloaded at the cells.

Suppose that the pressure at the lungs is P_L and the pressure at the cell is P_C. The amount of loading and unloading of oxygen can be read from the binding isotherm. In Fig. 5.1 we show three different binding isotherms, i.e. θ as a function of the pressure of oxygen. In a linear curve, the amount that is transported by the H_b molecule is proportional to the difference $\Delta\theta$ (i.e. the difference between the average number of oxygen molecules loaded at pressure P_L, and the average number of oxygen molecules which remain on the H_b at the pressure P_C). Ideally, we would like to have all the binding sites full at the loading end, and empty at the unloading end. This can be achieved when P_L is infinity and P_C is zero pressure. In reality, we are given fixed values of P_L and P_C, and we are interested in the maximum value of $\Delta\theta$, which is proportional to the maximum amount of oxygen transported between the two sites for each round-trip of the H_b molecule.

In Fig. 5.1a we see that in a typical binding isotherm (called Langmuir isotherm, see Sect. 5.1) the difference $\Delta\theta$ is not very large. The fraction θ at the lung is about 0.9, and the fraction at the cell is about 0.7. This means that only a small amount of oxygen is transported per one round trip of the H_b. For the linear curve in Fig. 5.1b, $\Delta\theta$ is a little larger, than in Fig. 5.1a.

Now suppose that instead of the curves in Fig. 5.1a or b the binding isotherm has the form as in Fig. 5.1c. Here, we see that the difference $\Delta\theta$ is quite large. This means that a large number of oxygen molecules will be transported per round trip on the H_b.

How does one get such a binding curve? Here is the secret; it is called *cooperativity* in the biochemical literature. In this chapter we call it *communication* between ligands, which is somewhat a more general term. How do ligands cooperate or communicate? The simple answer is by *direct* intermolecular interactions. A ligand in site 1, if it interacts favorably with site 2 causes an increase in the probability of binding of ligand to site 2. This, in turn increases the probability of binding to site 3, and this increases the probability of binding to site 4. We say that the ligands bind cooperatively. In the language of IT we can say that ligands communicate between themselves. Whatever the language we use, the effect of this communication is that when the first oxygen molecule binds to H_b it "signals" to all the other sites that the likelihood of binding to the second site and so on is enhanced. At the same time, when the pressure is reduced at the unloading site, once the first oxygen molecule is downloaded, it enhances the likelihood of downloading of the other oxygen molecules. The net effect is that $\Delta\theta$ (proportional to the average molecules transported per one round trip of the H_b molecules between the lungs and the cells), is quite large, as we see in Fig. 5.1c.

This mechanism of the cooperative binding was well understood for over a hundred years Adair [2], Pauling [3], Antonini et al. [4], Imai [5], Ben-Naim [1].

Indeed, the *interaction energy* between ligands occupying different binding sites is a possible explanation for the phenomenon of cooperativity. However, Pauling [3] was still puzzled by the outstanding efficient transport of oxygen by H_b. Pauling noticed that the distance between the binding sites in H_b is quite large (about 30 Å). At such distance the interaction energy between two oxygen molecules occupying two sites is negligible. It follows that although direct interaction energy can produce cooperativity, it cannot be the main cause for the efficient transport of oxygen by H_b. Pauling tried to explain the cooperativity of H_b using different interaction energies and different geometries of the H_b molecules. (For details about the mathematical Pauling's treatment of this phenomenon, see Ben-Naim [1]).

A major development in our understanding the way H_b operates was advanced by Monod et al. [6], and Monod et al. [7]. When I first read these two articles I felt that the authors were revealing to me one of cleverest and most intricate tricks of nature. It was one of the most exciting problems I had worked on which led me to the writing of a book on "Cooperativity and Regulation in Biochemical Processes," in 2001, [1].

The general idea of the Monod-Wyman-Changeux (MWC) model is quite simple to describe qualitatively. The mathematical treatment is described in detail in Ben-Naim [1]. The H_b molecule is known to be a tetramer molecule. It consists of four subunits roughly forming a regular tetrahedron. Each subunit contains one binding site for an oxygen molecule. When a ligand (here O_2) binds to one site, it causes a conformational change in the subunits on which the site is. This, in turn, changes the conformation of the neighboring subunits, which in turn causes a change in the neighboring sites. Thus, although there is no direct interaction between the ligands on the sites, there is an indirect communication. Thus, if the probability of binding to an empty H_b molecule is P_1, the conditional probability of binding to the second site, given the first site is occupied is $P_2 = P(2|1) > P_1$. The net effect is that once a site is occupied it "sends a signal" to all other sites. The signal is not sent by *direct interaction* but by *indirect conformational induced change* in the subunits. This indirect communication between the sites is the cause of the efficient transport of oxygen by H_b. We shall discuss this phenomenon in Sect. 5.5 in terms of mutual information (which includes the concept of cooperativity between ligands).

The effect of indirect communication is known as the *allosteric* effect (from ancient Greek, *allos* = other, and *stereos* = solid object). It means that a ligand causes change at a site other than the binding sites on which it is bound.

The allosteric model for H_b is truly a revolutionary idea in understanding the origin of cooperativity of binding and transporting of oxygen by H_b. It turns out that the same mechanism explains another, no less puzzling phenomenon known as enzymatic regulation. We shall briefly describe this phenomenon here (for more details, see Ben-Naim [1]).

Suppose that there is a sequence of chemical reactions; A produces B, B produces C, and C produces D. Suppose that each of this reaction is catalyzed by a specific enzyme $E(A \to B)$, $E(B \to C)$ and $E(C \to D)$ for the first, second, and the third reaction, respectively. We write this sequence of reaction as:

$$A \xrightarrow{E(A\to B)} B \xrightarrow{E(B\to C)} C \xrightarrow{E(C\to D)} D$$

We know that the concentrations of many compounds in our body is maintained within a very narrow range. For instance, when we measure the concentration of glucose in the blood we always find that it falls between two numbers, the minimum and the maximum. This is considered to be the "normal" range of concentrations of glucose. The question is: How does the body maintain this level of concentration of glucose, independently of how much glucose is introduced into our body by the food intake? The answer to this question is, "allosteric regulation." The mechanism of such regulation is similar to the one we have described for H_b. Here is how this mechanism works. Suppose that the compound D must be maintained with very narrow range of concentrations, say, between C_1 and C_2 (these two concentrations play the same role played by the two pressures in the case of H_b).

We know that enzymes work by first binding a *reactant* on an active site, and then the reactant is converted to a product, which is released from the binding process. In the series of reactions shown in the diagram above, suppose that initially there is an access of reactant A, this will produce more of the product B and this produces more of C, etc.

Now let us look at the sequence of reactions $A \to B \to C \to D$, and assume that the body requires a constant level of concentration of the product D. The amazing trick that evolved in nature is as follows: Suppose that the first enzyme $E(A \to B)$ can bind the product D at some binding site, which is different from the active site. This is called the regulatory site. In addition, suppose that the binding of D on the regulatory site causes a change in the conformation of the enzyme $E(A \rightarrow B)$, in such a way that the active site changes its structure and becomes less active.

In this schematic process the binding of D, on the *regulatory* site makes the active site less active, or not active at all. We can express the extent of activity of the enzyme as a function of the concentration of D.

Thus, when the concentration of D increases, the activity of the enzyme reduces, and therefore less D will be produced. Once the concentration of D increases beyond the level C_2, the enzyme $E(A \rightarrow B)$ will be deactivated. Therefore, this enzyme stops the production of B, which stops the production of C, which stops the production of D. The production of D will only resume when its concentration drops below C_1.

Thus, we see that the mechanism of allosteric regulation is essentially the same as the cooperative binding of oxygen to H_b. In both cases the binding of a ligand induces a conformational change in the adsorbent molecule, which causes a change in the properties of another allosteric site. For mathematical details see Ben-Naim [1].

In the following sections of this chapter we shall proceed from the simplest system of molecules with one binding site, to more complicated systems of two, three, and four-site system. We will also study systems with direct ligand-ligand interaction as well as systems with only indirect interactions.

5.1 The Method of Generating Probabilities

In this section, we describe very briefly the procedure of calculating the probabilities of various events. The theory underlining this procedure is statistical mechanics. The reader who is not familiar with statistical mechanics can skip this section and go directly to the next section where we shall use these probabilities to calculate SMIs and various MI relevant to the binding systems. More details about the statistical mechanical procedure may be found in references [1, 8, 9].

The basic system will be an adsorbent molecule usually a protein (**P**) having a number of binding sites (m) to some ligand **A**.

The adsorbent molecule **P** may be in different conformations, and the adsorbing site could be either empty or occupied.

As an example suppose that we have a protein **P**, which consists of two subunits. Each subunit can be in one of two states; L and H (for lower and higher energy). Therefore, the adsorbent molecule can be in one of the four states shown in Fig. 5.2. The four states of an empty adsorbent molecule are: LL, LH, HL, and LL. Each of sites of the adsorbent molecule may be either empty or occupied by a ligand A. Therefore, the adsorbent molecule can be in one of the sixteen states shown in Fig. 5.3. Upper panel shows the four states of an empty adsorbent molecule, and the lower panels show the various states of occupancy of the adsorbent molecules.

The fundamental quantity from which we generate the probabilities is the grand partition function (GPF) of a single adsorbent molecule, which is written as:

Fig. 5.2 Four possible states of a two-subunit system, each can be in one of two conformers: L or H, represented by a square and a circle, respectively

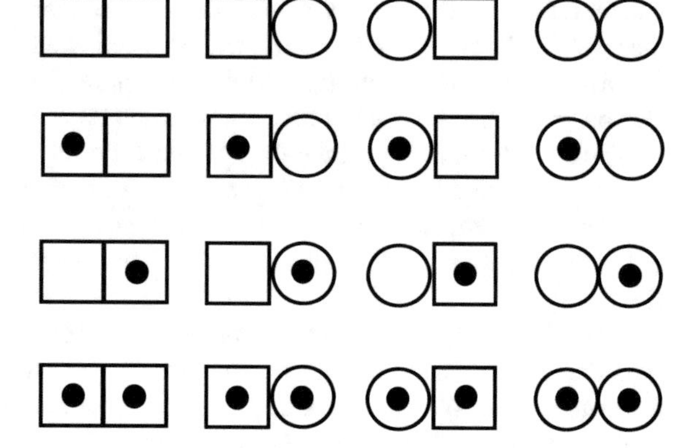

Fig. 5.3 All sixteen states of a two-subunit system. Each subunit can be in one of two conformers: Either L or H, and can be either empty or occupied

$$\xi = \sum_{i=1}^{2} Q(i)\lambda^{i} \tag{5.1}$$

where $Q(i)$ is the canonical PF of the adsorbent molecule having i ligands, and λ is the absolute activity of the ligand.

The canonical partition function (PF) $Q(i)$ is constructed from all possible configurations of the system having i ligands. For example $Q(2)$ is the sum over four terms, see lower panel in Fig. 5.3.

$$Q(2) = \sum_{\alpha\beta} Q_{\alpha} Q_{\beta} Q_{\alpha\beta} q_{\alpha} q_{\beta} q_{\alpha\beta} \tag{5.2}$$

where we define the following quantities:

$$\begin{aligned} Q_{\alpha} &= \exp[-E_{\alpha}/k_B T] \\ Q_{\alpha\beta} &= \exp\left[-E_{\alpha\beta}/k_B T\right] \\ q_{\alpha} &= \exp[-U_{\alpha}/k_B T] \\ q_{\alpha\beta} &= \exp\left[-U_{\alpha\beta}/k_B T\right] \end{aligned} \tag{5.3}$$

Thus, each term in Eq. (5.2) contains six factors, Q_{α} depending on the energy state E_{α} of the first subunit (α and β may either be in L or H). $Q_{\alpha\beta}$ depends on the interaction energy between two subunits ($E_{\alpha\beta}$), when one is in state α, and the second is in state β. q_{α} depends on the binding energy (U_{α}) of the ligand to the site on the subunit being in state α, and $q_{\alpha\beta}$ depends on the interaction energy ($U_{\alpha\beta}$) between the two ligands occupying the sites on the two subunits, being in states (α and β). The sum in Eq. (5.2) is over all states of α and β, i.e. α and β may be either L or H.

The ligand **A** is assumed to be at equilibrium with either a gaseous phase or a liquid phase containing the ligand at some concentration C_A. The absolute activity (λ) of the ligand is related to chemical potential by:

$$\lambda = \exp[\mu/k_B T] \tag{5.4}$$

For all intents and purposes we shall assume that the absolute activity is proportional to the concentration of the ligand in the gaseous (or the liquid) phase, i.e.

$$\lambda = \lambda_0 C_A \tag{5.5}$$

We shall examine the dependence of the SMI and the MI on both the temperature T and ligand concentration C_A. The Boltzmann constant will be taken as unity ($k_B = 1$).

To generate the probabilities, we simply take the relevant term in the GPF and divide it by the GPF. For instance, if we are interested in the probability of finding the adsorbent molecule empty in a state LH we simply take the ratio:

$$P(L, 0, H, 0) = \frac{Q_L Q_H Q_{HL} \lambda^0}{\xi} \tag{5.6}$$

Or, if we want the probability that the adsorbent molecule is fully occupied and in state HH, we take the ratio:

$$P(H, 1, H, 1) = \frac{Q_H Q_H Q_{HH} q_H q_H q_{HH} \lambda^2}{\xi} \tag{5.7}$$

Note that we use 0 and 1 to describe an empty or occupied site respectively. The meaning of these probabilities is as follows. We think of an ensemble of many adsorbent molecules. In such an ensemble we can interpret the probabilities as mole fractions of the adsorbing molecules in the specific state.

In studying adsorbing of ligands on proteins we are interested in the average occupancy of the protein. This average is defined by:

$$\overline{n} = \sum_{i=1}^{m} i P(i) \tag{5.8}$$

where $P(i)$ is the probability of finding the adsorbent molecules with i ligands, and the sum is over all possible i from one to m. This average can also be determined from the GPF by:

$$\overline{n} = \lambda \left(\frac{\partial \ln \xi}{\partial \lambda} \right)_{\Gamma} \tag{5.9}$$

The dependence of $\overline{n}$ on λ (or on the ligand concentration C_A) is referred to as the *binding isotherm*. Sometimes, one is interested in the average occupancy per one site. This is defined by:

$$\theta = \frac{\overline{n}}{m} \tag{5.10}$$

The function $\theta(\lambda)$ is important in applications of the theory to many biological systems, such as the efficiency of transporting of oxygen by hemoglobin. We shall discuss these systems in the following sections of this chapter.

5.2 Adsorbing on a Single-Site Molecule

Although the case of a single-site adsorbing molecule does not exhibit any transmission of information, it is important to get familiar with such a system in order to understand the more complicated systems with two or more binding sites.

We assume that the adsorbent molecule P can be in two states L and H, and each molecule has one binding site, which can either be empty (0), or occupied (1). Figure 5.4 shows all four possible states of this molecule.

The GPF of this system is:

$$\xi(T, \lambda) = Q_L + Q_H + (Q_L q_L + Q_H q_H)\lambda \tag{5.11}$$

The first two terms on the right-hand-side of (5.11) pertain to the two empty sites. The third term pertains to the two occupied states of the adsorbent molecule. The probabilities of the empty molecules are:

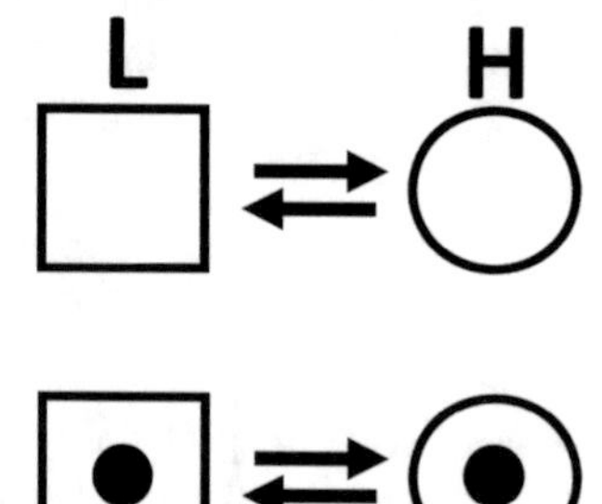

Fig. 5.4 The four possible states of a single-site adsorbent molecule; empty L, empty H, occupied L, and occupied H

$$\Pr(L, 0) = \frac{Q_L}{\xi} \tag{5.12}$$

$$\Pr(H, 0) = \frac{Q_H}{\xi} \tag{5.13}$$

When we have an ensemble of adsorbent molecules, we may interpret these probabilities in (5.12) and (5.13) as the mole fractions of empty molecules in state L and H, respectively.

Similarly, we can define the probabilities of the occupied states of the molecule:

$$\Pr(L, 1) = \frac{Q_L q_L \lambda}{\xi} \tag{5.14}$$

$$\Pr(H, 1) = \frac{Q_H q_H \lambda}{\xi} \tag{5.15}$$

We can also define the probabilities of L and H, regardless of the state of occupancy, these are:

$$\Pr(L) = \frac{Q_L + Q_L q_L \lambda}{\xi} \tag{5.16}$$

$$\Pr(H) = \frac{Q_H + Q_H q_H \lambda}{\xi} \tag{5.17}$$

These are also the mole fractions of molecules in state L and H. We also define the equilibrium constant, for the conversion between L and H, by:

$$K(\lambda) = \frac{\Pr(H)}{\Pr(L)} = \frac{Q_H + Q_H q_H \lambda}{Q_L + Q_L q_L \lambda} \tag{5.18}$$

Clearly, this equilibrium constant depends on the ligand concentration through λ. When $\lambda = 0$, we have empty molecules, and the equilibrium constant is:

$$K(0) = \frac{Q_H}{Q_L} \tag{5.19}$$

When $\lambda \to \infty$, we have the equilibrium constant of occupied molecules:

$$K(1) = \frac{Q_H q_H}{Q_L q_L} \tag{5.20}$$

We define the two quantities:

$$K = \frac{Q_H}{Q_L}, \quad h = \frac{q_H}{q_L} \tag{5.21}$$

K is a measure of the relative concentrations of the two conformations at $\lambda = 0$, and h is a measure of the relative strength of the binding of the ligand to the two conformations. In the following we shall study the various SMI as a function of the parameters λ, K and h.

In Eq. (5.10), we defined the fraction θ. An important quantity of interest in the theory of binding of ligands to biopolymers is the effect of the ligand on the probability $\Pr(L)$, or equivalently, the effect of binding a ligand on the concentration of the L conformation. One way to express this effect is by the quantity:

$$d_L = \frac{\partial \Pr(L)}{\partial \theta} = \frac{K(1-h)}{(1+K)(1+hk)} \tag{5.22}$$

It is clear from Eq. (5.22) that, when either $K = 0$ (i.e. L is infinitely more stable than H), or when $h = 1$ (i.e. the ligand has no preferential binding to the two conformations), there will be no effect of the ligand on the concentration of L. In this case we say that the binding of the ligand does not induce a conformational change in the protein P. When h is different from unity, say, $h > 1$, the binding energy of the ligand to H is stronger than to L, for instance see Fig. 5.5. We shall see in the following sections that conformational induced change in the protein is essential in understanding the cooperativity between ligands, or in the extent of "communication" between ligands occupying different sites.

Finally, for calculating SMI we shall use the probability distribution:

$$\Pr(0) = \frac{Q_L + Q_H}{\xi} \tag{5.23}$$

$$\Pr(1) = \frac{Q_L q_L + Q_H q_H}{\xi} \tag{5.24}$$

These are the probabilities of finding the site empty or occupied, independently of the conformational state of the protein.

In the following we present a few results pertaining to this system. Understanding this system is crucial to understanding the phenomenon of cooperativity and transmission of information between different sites.

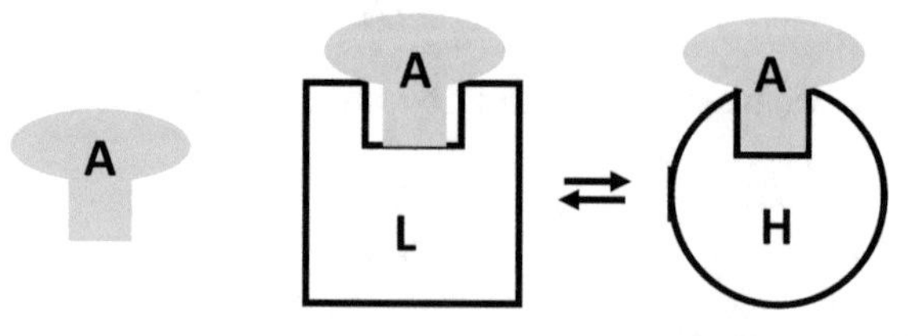

Fig. 5.5 The binding energy of the ligand A might be different to the sites of the L and the H conformers of the adsorbing molecule

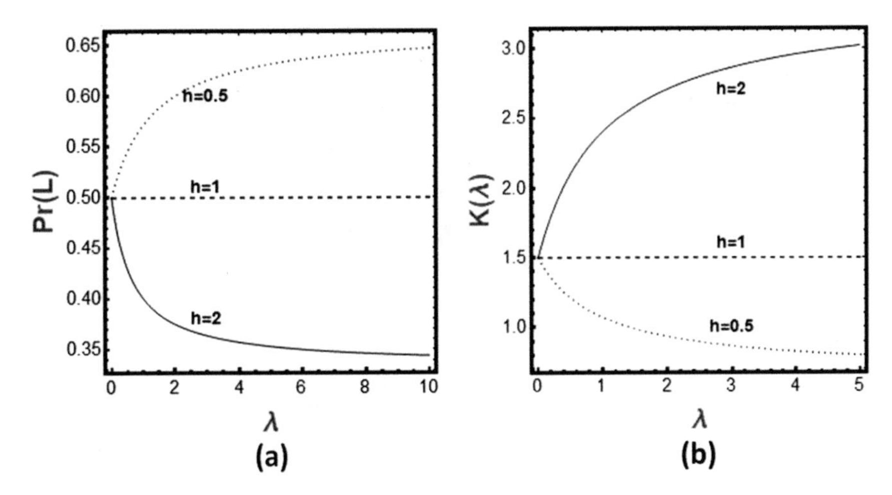

Fig. 5.6 **a** The mole fractions of L as a function the absolute activity λ of the ligand. **b** The dependence of the equilibrium constant on the absolute activity of the ligand

Figure 5.6a shows the probability of finding the molecule in state L, regardless of its state of occupancy as a function of λ (which is basically the concentration of the ligand), for $K = 1$, and three values of h. $h = 1$ means that the ligand has no preference to bind to either L, or H. The choice of $K = 1$, means that L and H are initially ($\lambda = 0$) equally probable; $\Pr(L) = \Pr(H) = 0.5$. Increasing λ will increase the binding on the adsorbent molecule, but will not change the relative concentrations of L and H.

When $h = 2$, meaning that the ligand prefers binding to H. Therefore, it will increase the concentration of H, which is equivalent to the decrease of the concentration of L. The opposite effect is seen when $h = 0.5$, meaning that the ligand prefers binding to L.

Figure 5.6b shows the change in the equilibrium constant defined in Eq. (5.18). $h > 1$ means preference for H, hence the equilibrium constant increases with λ. The opposite effect occurs when $h < 1$.

Figure 5.7 shows a typical binding isotherm. $\theta(\lambda)$ is the fraction of sites occupied as a function of λ. When $\lambda = 0$, there is no ligand binding, hence, $\theta = 0$. When λ becomes very large, all the sites will eventually be occupied, and $\theta \to 1$. We see that all the curves in this figure are similar. This type of curve is called Langmuir isotherm. Later on in this chapter we shall see that the binding isotherm of cooperative system could be very different from the one shown in Fig. 5.7. As we noted in the introduction to this chapter the form of the binding isotherm is the key to understanding the efficiency of transporting of oxygen by hemoglobin as well as the efficiency of regulatory enzymes.

Figure 5.8 shows how the derivative defined in Eq. (5.22) changes with K for various values of h. When $K = 0$ (or infinity), the L form is most stable, and all the curve start at $d_L = 0$. When K increases, d_L also increases until it reaches a

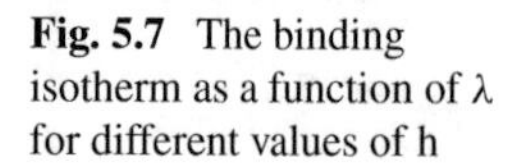

Fig. 5.7 The binding isotherm as a function of λ for different values of h

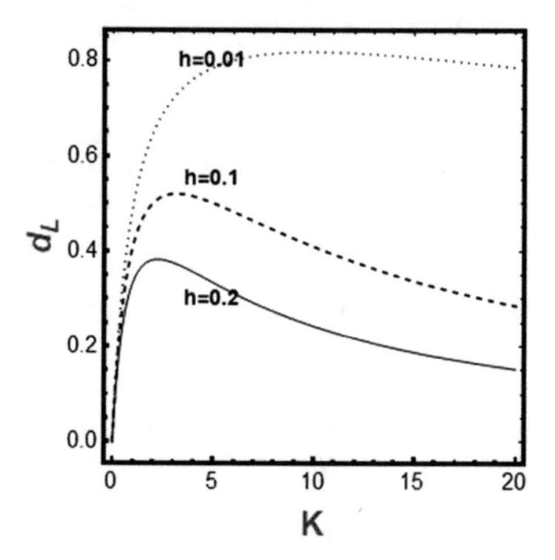

Fig. 5.8 The differential conformational change induced by the ligand, as a function of K for different values of h

maximum. The maximum occurs when $K = \sqrt{1/h}$. When $K \to \infty$, all the curves tend to zero, meaning that there is no induced conformational change.

Figure 5.9a shows the behavior Pr(0) and Pr(1), and the SMI defined on this distribution for $K = 1, h = 1$. (Similar curves are obtained for other values of K and h). Clearly, when λ is zero, all the sites are empty; $\Pr(0) = 1$ and $\Pr(1) = 0$. When $\lambda \to \infty$ all the sites are occupied; $\Pr(0) = 0$ and $\Pr(1) = 1$. These two curves cross at the point when $\Pr(0) = \Pr(1) = 0.5$. Corresponding to this behavior of the distribution $(\Pr(0), \Pr(1))$, the SMI starts at zero for $\lambda = 0$ (meaning a distribution $(1, 0)$), and end up at zero for $\lambda \to \infty$ (meaning a distribution $(0, 1)$). As we increase λ, there is a point at which the distribution is $(0.5, 0.5)$ for which SMI $= 1$, Fig. 5.9b.

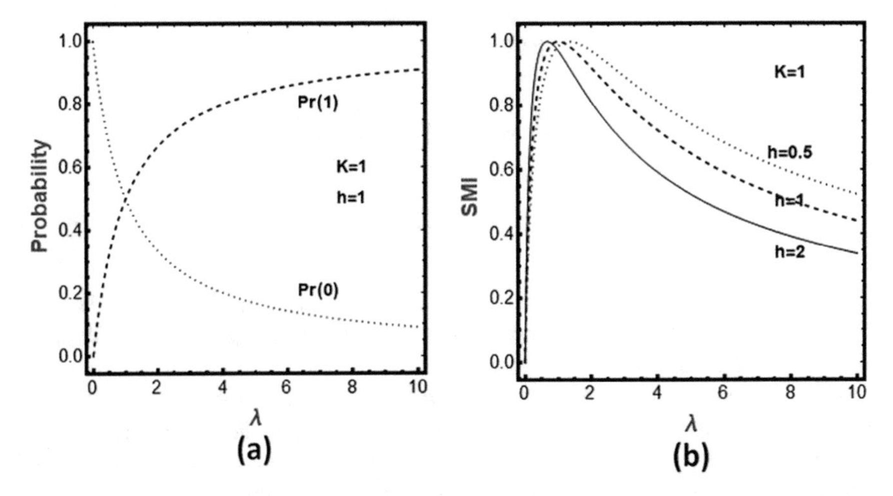

Fig. 5.9 **a** The singlet probabilities as a function of λ, for K = 1 and h = 1. **b** The singlet-SMI as a function of λ for different values of h

5.3 Two-Site Systems

In this section, we discuss two-site systems. We have a macromolecule, usually a protein which we denote **P**, which has two sites, on each of which a ligand A can be adsorbed. We shall always assume that the two sites are equivalent, and only one type of ligand can occupy the sites. We shall denote the two sites by 1 and 2. More general cases are discussed in Ben-Naim [1].

We discuss separately three kinds of *communication* between the two sites; first, due to *direct* interaction between ligands occupying the two sites; second, *indirect* communication by means of a change of conformation of the entire adsorbent molecule; and third, an *indirect* communication by means of changes in the conformation of each subunit of the protein.

As usual, all the relevant probabilities and correlations are derived from statistical mechanics. The reader who is not familiar with statistical mechanics can skip the rest of this section and proceed to Sect. 5.3.1 where we discuss the probabilities which we shall use in calculating SMIs and MIs.

We start with the fundamental quantity, the GPF of a single adsorbent molecule defied by:

$$\xi = Q(0,0) + [Q(0,1) + Q(1,0)]\lambda + Q(1,1)\lambda^2 \tag{5.25}$$

The "events" here are symbolically written as (0, 0), (0, 1), (1, 0) and (1, 1), meaning: empty-empty, empty-occupied, occupied-empty and occupied-occupied, respectively. The corresponding probabilities are:

$$\Pr(0,0) = \frac{Q(0,0)}{\xi} \tag{5.26}$$

$$\Pr(1,0) = \Pr(0,1) = \frac{Q(1,0)\lambda}{\xi} \tag{5.27}$$

$$\Pr(1,1) = \frac{Q(1,1)\lambda^2}{\xi} \tag{5.28}$$

$\Pr(0,0)$ is the probability that both sites are empty. $\Pr(1,0)$ is the probability that site 1 is occupied and site 2 is empty. (This is the same as $\Pr(0,1)$). $\Pr(1,1)$ is the probability that both sites are occupied.

It should be noted that there are three different *singlet* probabilities: The first, $\Pr(1,0)$, is the probability that the first site is occupied, while the second site is empty. The second, is the probability that the molecule **P** is *singly occupied* irrespective of which site is occupied, is denoted by $\Pr(1)$, and is defined by:

$$\Pr(1) = \Pr(0,1) + \Pr(1,0) \tag{5.29}$$

The third, and the more important singlet probability, is defined by:

$$\Pr(1,\_) = \Pr(1,0) + \Pr(1,1) \tag{5.30}$$

This is the probability that the first site is occupied when the second site's state is unspecified; it can either be empty or occupied. Similarly, we define the probability:

$$\Pr(0,\_) = \Pr(0,0) + \Pr(0,1) \tag{5.31}$$

This is the probability of finding the first site empty, regardless of the state of occupancy of the second site. These probabilities are shown in Fig. 5.10.

As we discussed in Sect. 5.2, the binding isotherm per site is obtained by:

$$\theta = \frac{\overline{n}}{m} = \frac{\lambda}{m}\frac{\partial \ln \xi}{\partial \lambda} \tag{5.32}$$

In our case, $m = 2$, the maximum of $\overline{n}$ is 2, hence, θ varies between zero to 1.

There are four correlations in these system, which we shall use to calculate the mutual information (MI), these are:

$$g(0,0) = \frac{\Pr(0,0)}{\Pr(0,\_)^2} \tag{5.33}$$

$$g(1,0) = g(0,1) = \frac{\Pr(1,0)}{\Pr(0,\_)\Pr(1,\_)} \tag{5.34}$$

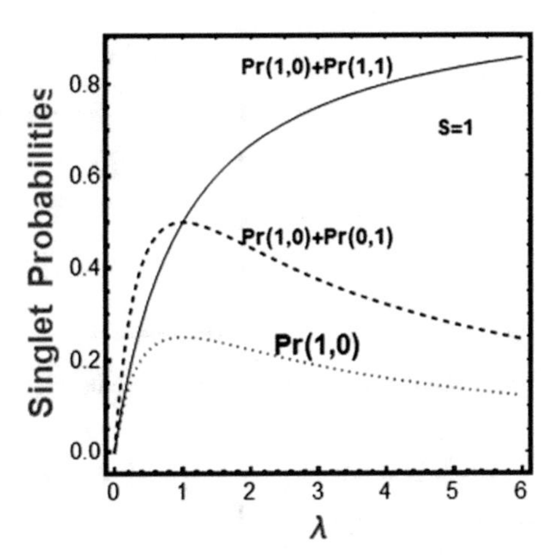

Fig. 5.10 Three different singlet probabilities as a function of λ, for S = 1

$$g(1,1)=\frac{\Pr(1,1)}{\Pr(1,\_)^2} \tag{5.35}$$

In the study of cooperativity of binding, only the correlation between the two ligands occupying both sites, is of importance. In this chapter, we shall need all of the four correlation functions in order to calculate MI. Note that all these correlations are dependent on λ (recall that λ is essentially the concentration of the ligand in the gaseous or the solution phase). In the theory of binding isotherm only the limit $\lambda \to 0$ of Eq. (5.35) is used. It should be noted that in this limit both the numerator and the denominator in Eq. (5.35) tend to zero. However, the ratio tends to a constant and measures the extent of cooperativity between the two ligands. In the following sections we shall study the correlations and their dependence on λ in connection with the MI between the two sites.

5.3.1 Direct Communication via Ligand-Ligand Interaction

The first and the simplest model of a two-site system with communication is shown in Fig. 5.11. Here we have two equivalent sites. When the two sites are occupied the interaction energy between the two ligands is $U(1,2)$. The corresponding GPF for this case, Eq. (5.25) reduces to:

$$\xi = 1 + (q+q)\lambda + q^2 S\lambda^2 \tag{5.36}$$

where $q = \exp[-U/k_BT]$, with U the ligand-site interactions, and $S = \exp\left[-\frac{U(1,2)}{k_BT}\right]$, where $U(1,2)$ is the ligand-ligand interaction (indicated by the double arrow in Fig. 5.11).

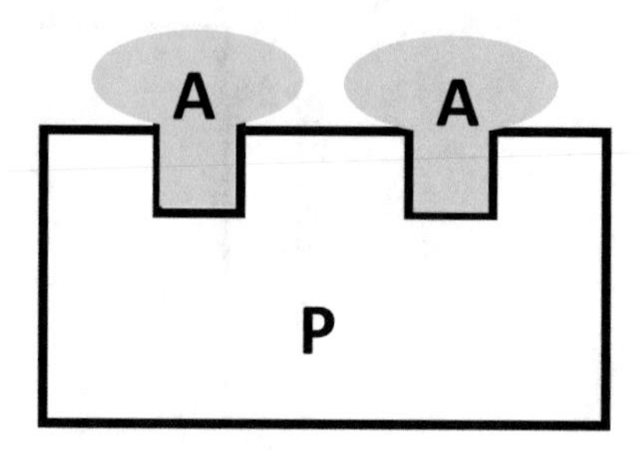

Fig. 5.11 The model of a two-site molecule with direct interactions between the two ligands

Whenever, $U(1,2) < 0$, we get $S > 1$, and we say that there is *positive* cooperativity. When $U(1,2) > 0$, we have $S < 1$, and we say there is *negative* correlation or negative cooperativity between the two sites.

As we noted above there are three singlet probabilities. These are shown in Fig. 5.10, for $S = 1$. $\Pr(1,0)$ starts at zero for $\lambda = 0$. Once we increase the concentration of the ligand we "pump" in more ligand on the adsorbent molecule, and therefore $\Pr(1,0)$ starts to increase. This function must go through a maximum then drop to zero for $\lambda \to \infty$. This is so because at very large concentration all the adsorbent molecules will be doubly occupied; hence, $\Pr(1,1)$ tends to one, but $\Pr(1,0)$ must tend to zero.

The second is $\Pr(1,0)+\Pr(0,1)$ is simply twice the value of $\Pr(1,0)$ (for identical sites), therefore, it should have the same behavior as that of $\Pr(1,0)$.

The third curve clearly tends to one at $\lambda \to \infty$, because at very high concentrations of the ligand both sites will be occupied, hence $\Pr(1,1)$ tends to one.

Figure 5.12 shows the dependence of $\Pr(0,\_)$ and $\Pr(1,\_)$ on λ for different values of S. Clearly, the larger the interaction between the two ligands the steeper the increase of $\Pr(1,\_)$, and the decrease of $\Pr(0,\_)$.

Figure 5.13 shows the probabilities $\Pr(0,0)$, $\Pr(1,0)$, and $\Pr(1,1)$ as a function of λ for the values of S.

Figure 5.14 shows a typical binding isotherm for the two-site case with different values of S. All curves start at zero and tend to one for $\lambda \to \infty$. The larger S, the steeper the "filling-up" of the adsorbent molecule by ligands until reaching total saturation at $\theta = 1$.

Figure 5.15 shows the pair correlation function $g(1,1)$ for various values of S. This correlation is the most important one in the study of cooperativity of binding system. When $S = 1$, there is no correlation, and $g(1,1) = 1$. For $S < 1$, the correlation is also smaller than 1 (and $\log g(1,1)$ is negative), and for $S > 1$, $g(1,1)$ is also larger than one. Note that in all cases, when $\lambda \to 0$ the correlation $g(1,1)$ tends to the value of S. When $\lambda \to \infty$ all sites will be occupied with probability one. In this limit the correlation will tend to 1 (no correlation).

Figure 5.16 shows the correlation $g(1,0)$ and $g(0,0)$ which will be needed for calculating the MI, see Ben-Naim [10]. All the curves of $g(0,0)$ starts at $g(0,0) = 1$ for $\lambda = 0$, and tends to S when $\lambda \to \infty$. Note that both the numerator and

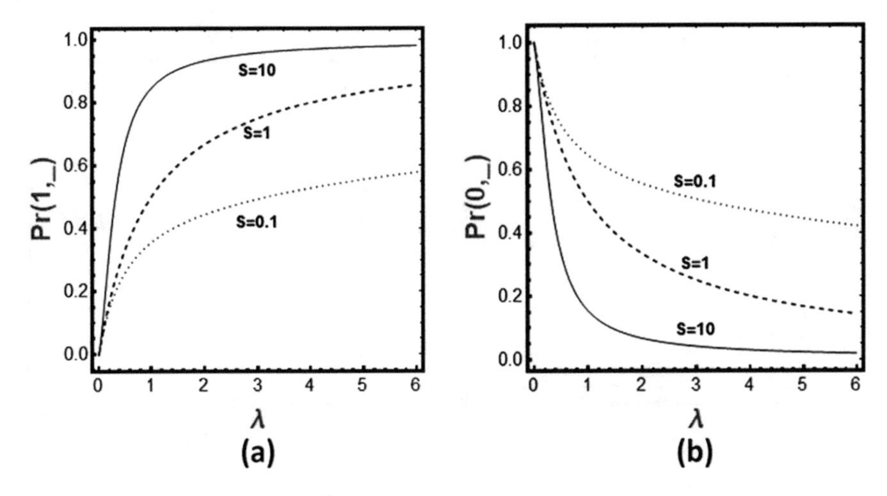

Fig. 5.12 The singlet probabilities as a function of **λ** for different values of S

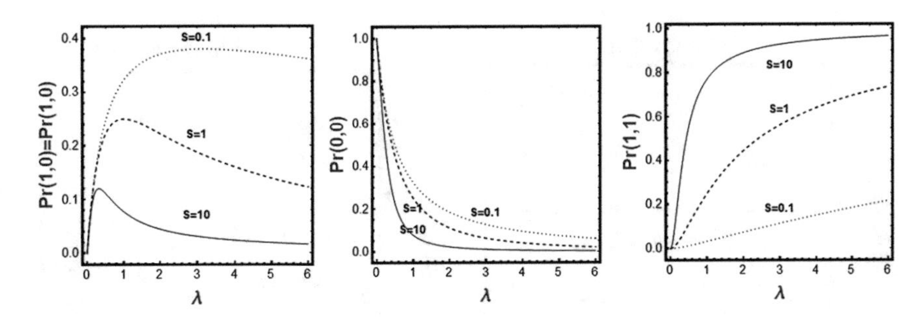

Fig. 5.13 The three pair-probabilities as a function of λ for different values of S

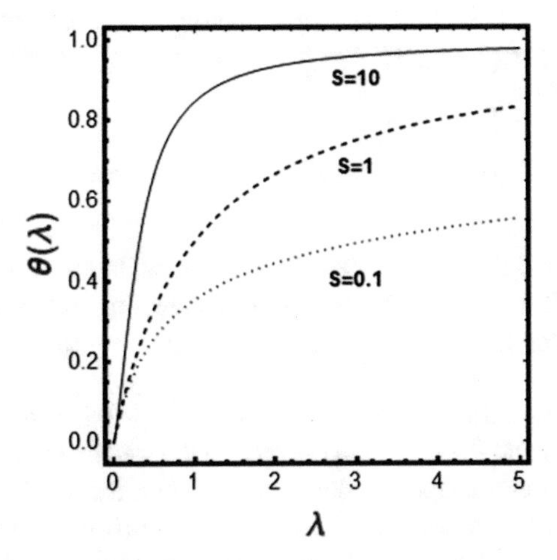

Fig. 5.14 The binding isotherm as a function of λ for different values of S

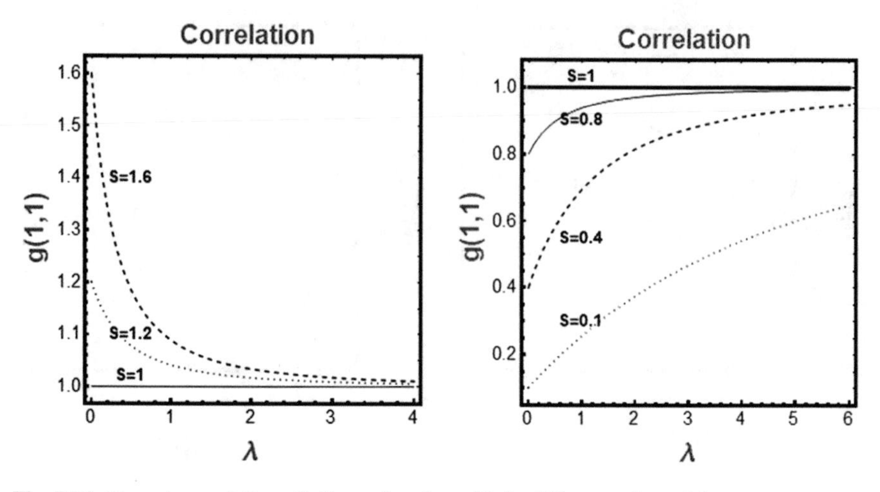

Fig. 5.15 The pair correlation g(1,1) as a function of λ for different values of S

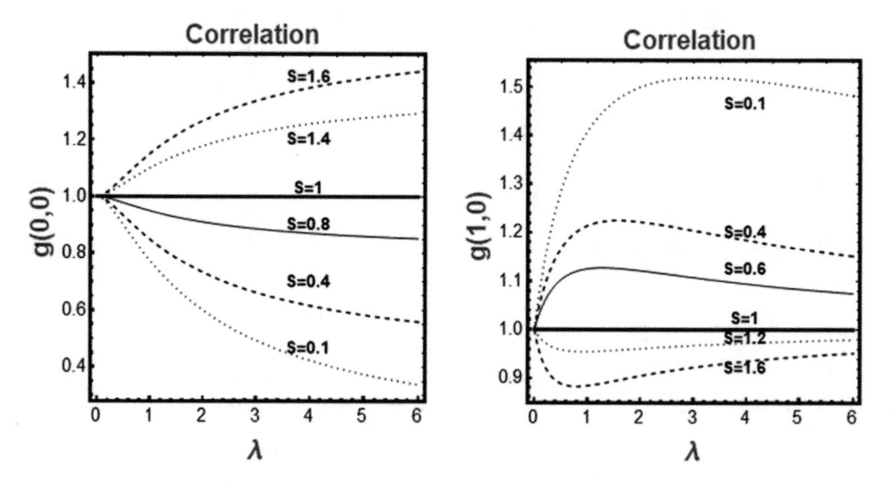

Fig. 5.16 The pair correlations, g(0,0) and g(1,0) as a function of λ for different values of S

denominator of Eq. (5.33) tends to zero when $\lambda \to \infty$, but the ratio, defining $g(0, 0)$ tends to S in this limit. A different behavior is shown by $g(1, 0)$. Again, all curves of $g(1, 0)$ start at one for $\lambda \to 0$, but in the limit of $\lambda \to \infty$, all the curves tend to 1, i.e. no correlation.

Figure 5.17 shows the two SMIs, the singlet SMI based on the probability distribution $\Pr(0, \_)$ and $\Pr(1, \_)$, and the pair SMI based on the distribution $\Pr(i, j)$.

Note that the general shapes of the curves of both the singlet and the pair SMI are quite similar. All curves start at zero for $\lambda \to 0$ (because the empty state has probability one), and all must tend to zero when $\lambda \to \infty$ (because the fully occupied sites have probability one). When λ increases from zero, there are finite probabilities

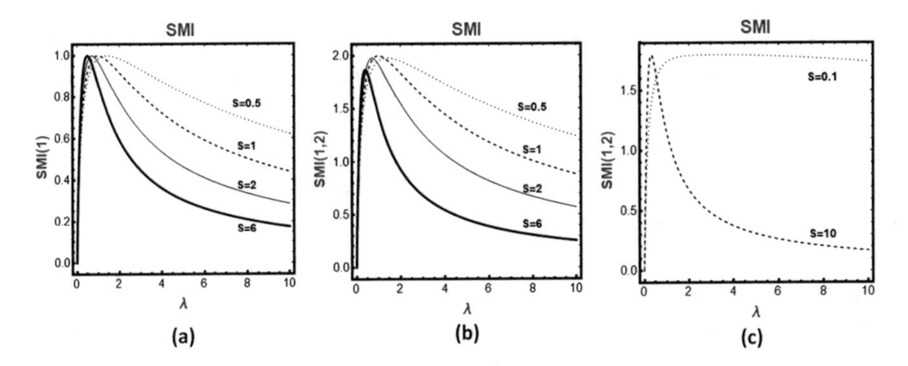

Fig. 5.17 The Singlet (**a**) and the pair-SMI (**b**) and (**c**) as a function of λ for different values of S

for all the states of occupancies, at some value of λ, the SMI reaches its maximal value; $\text{SMI}(1) \simeq 1$, and $\text{SMI}(1, 1) \approx 2$ (corresponding to two and four, nearly equal probability states of occupancy). Note however, that when S is too small or too big, the maximum of SMI(1, 2) does not reach the value of 2 (Fig. 5.17c). The reason is that for very strong interactions, either attractive or repulsive, there is no λ for which all the occupancy states are equally probable.

Finally, in Fig. 5.18 we show the mutual information for this system. As we expect when $S = 1$, there is no correlation and therefore the MI is zero, Ben-Naim [10]. All the curves tend to zero in either limits $\lambda \to 0$, or $\lambda \to \infty$. Note that the MI is always positive as it should be, independently of the individual pair correlations.

To conclude this subsection, we can say that all the results of the direct interactions are as expected. There are no surprises in this case. In general, stronger interactions lead to larger correlations (either positive or negative), and larger MI. Also, the

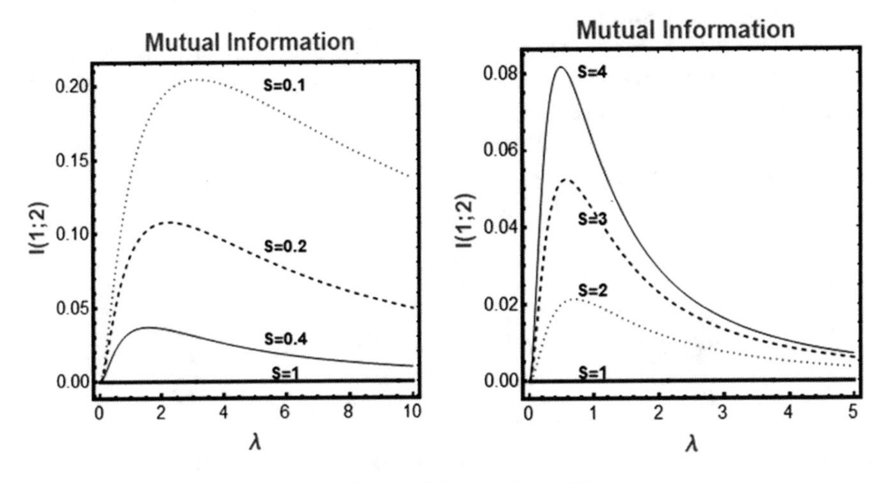

Fig. 5.18 The pair-MI as a function of λ for different values of S

dependence on λ is as expected. In the next two sections, we shall study two cases where the outcomes, although they make sense, are not easily expected.

5.3.2 Indirect Communication via Conformation Changes in the Adsorbent Conformation

The model in this case is shown in Fig. 5.19. The adsorbent molecule as a whole can be in one of two conformational states: L and H. Each molecule has two binding sites which have different shapes in the two conformations, such that the binding energy of the ligand A to the site L is different from the binding energy to the site in H, Fig. 5.19.

As in Sect. 5.3.1 the two conformations are at equilibrium. The equilibrium constant for the empty polymer, denoted K, is related to the energies of the two states of the two conformations:

$$K = \frac{Q_H}{Q_L} = \exp\left[-\frac{E_H - E_L}{k_B T}\right] \tag{5.37}$$

We denote the binding energy of the ligand to the two kinds of sites by U_L and U_H, and we define the ratio, h, as in Sect. 5.2, by:

$$h = \frac{q_H}{q_L} = \exp\left[-\frac{U_H - U_L}{k_B T}\right] \tag{5.38}$$

Because of the difference in the binding energies to the L and H sites, a ligand A binding to the molecule, will cause a shift in the equilibrium concentrations of L and H towards that conformation to which the ligand is bound stronger. In Fig. 5.19 we show that A "fits" better to the sites in L, and therefore we expect that whenever we "pump" ligands (increasing λ) into the system, the equilibrium will shift towards more of the L conformation as we have discussed in Sect. 5.2.

Figure 5.20 shows all eight states of the molecule; two empty, four singly occupied, and two doubly occupied.

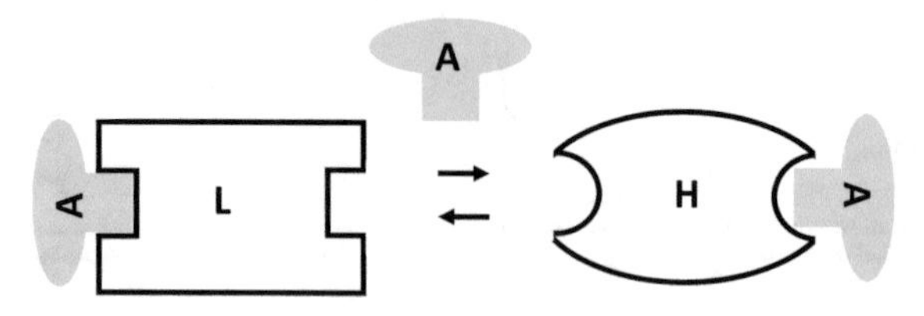

Fig. 5.19 The model of a two-site molecule having two conformations. The binding energies to the sites of the L and the H conformers are different

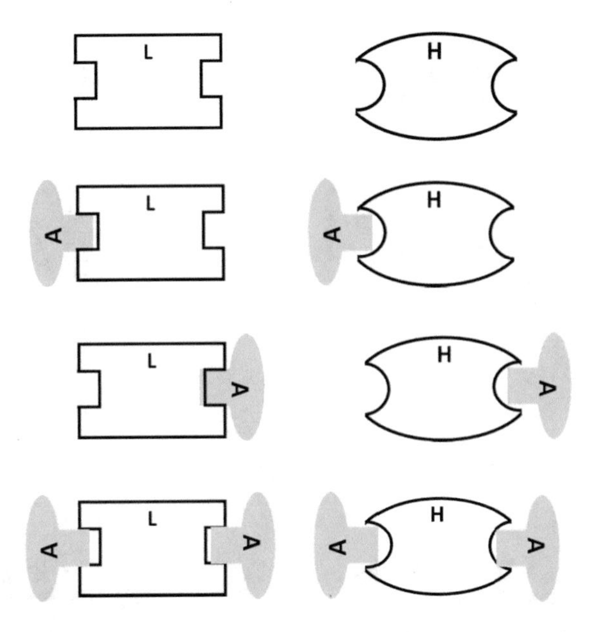

Fig. 5.20 All eight possible states of a two-site adsorbing molecule having two conformations

In Sect. 5.3.2, we studied the communication between the two sites due to direct interaction between the two ligands. Here, the means of communication is different. Suppose that we start with any initial concentrations C_L° and C_H° of the two conformers. When the first ligand binds to the molecule, it will most likely bind to the site L (to of which has stronger binding energy, Fig. 5.19). Therefore, it will shift the equilibrium constants towards higher concentration of the L conformation. This means that the next ligand approaching the molecule, will most likely encounter an L-site rather than an H-site. As we shall soon see this kind of correlation is always positive; binding on one site will always increase the probability of binding to the second site.

To highlight this kind of indirect correlation we shall assume that the two sites are far away from each other. Therefore, the direct interaction between the ligands is negligible and all the correlations (as well as cooperativity) are due to indirect effect.

The most surprising result is that the behavior of this system is quite similar to the system with direct interaction discussed in the previous Sect. 5.3.1.

We start with the binding isotherm which, as in the previous case is a typical Langmuir isotherm. Figure 5.21 shows a few binding isotherm for different values of h, and $K = 10$. Recall that the larger h, the larger the effect of the ligand on the conformational equilibrium. We also see that the larger h, the steeper the binding isotherm.

Figure 5.22 shows the three different singlet probabilities in this system. This should be compared with Fig. 5.10. The interpretation of these curves is the same as discussed in Sect. 3.5.1.

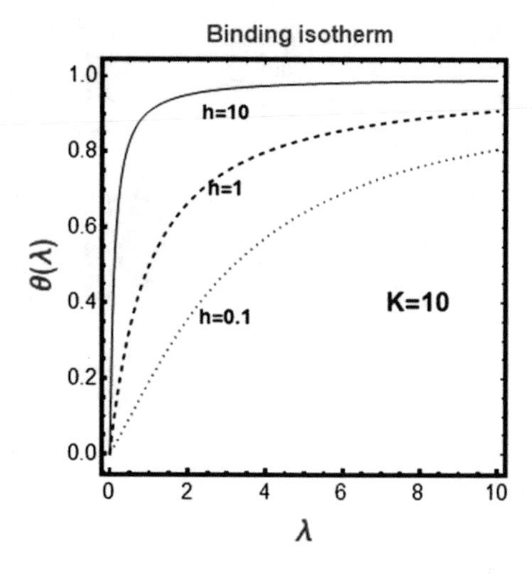

Fig. 5.21 The binding isotherm as a function of λ for K = 10 and different values of h

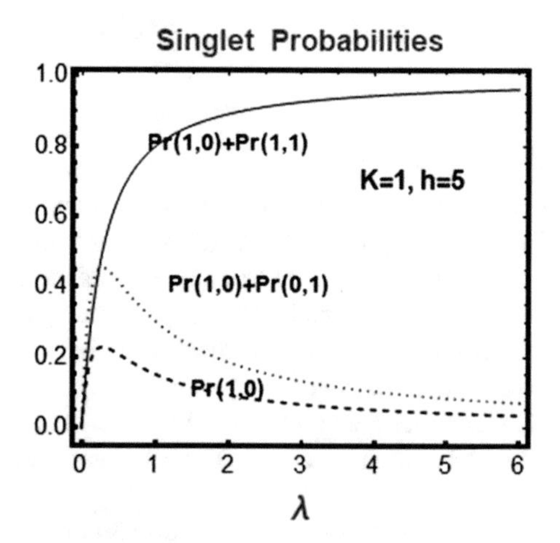

Fig. 5.22 Three different singlet probabilities as a function of λ, for K = 1 and h = 5

Figure 5.23 shows the dependence of the singlet distribution $\Pr(1, \_)$ and $\Pr(0, \_)$ as a function of λ for different values of h, and $K = 1$. Again, the general behavior is similar to the case of direct interactions discussed in Sect. 3.5.1, and Fig. 5.12.

Before we discuss the different pair correlations in this system, it should be noted that the only correlations which enter into the binding isotherm, and is related to the cooperativity in the system is the limiting value of the correlation:

$$g^0(1,1) = \lim_{\lambda \to 0} g(1,1) = \frac{(1+K)\left(1+Kh^2\right)}{(1+Kh)^2} = 1 + \frac{K(1-h)^2}{(1+Kh)^2} \tag{5.39}$$

where $g(1, 1)$ was defined in Eq. (3.35). This correlation is purely indirect correlation between the two events: "site 1 is occupied," and "site 2 is occupied". As it is seen

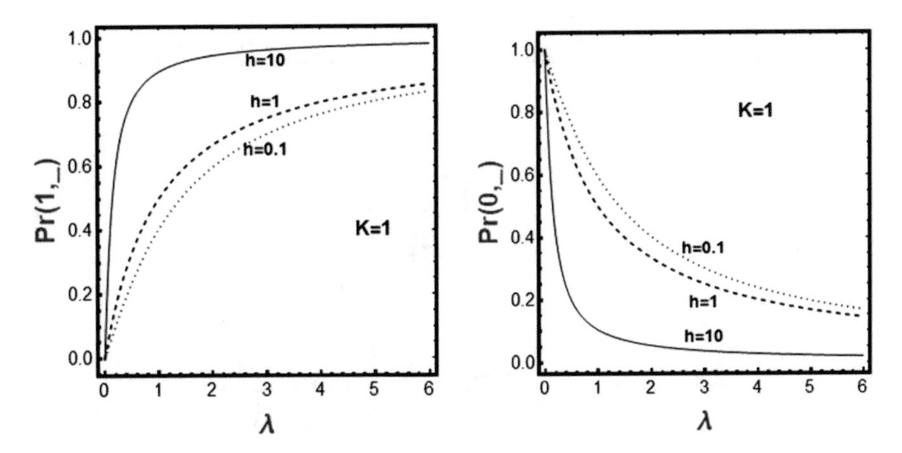

Fig. 5.23 The singlet probabilities as a function of λ, for K = 1 and different values of h

from the last term in Eq. (5.39), the correlation is always larger than one. It is equal to one, when either $h = 1$ (i.e. when the ligand does not have preference to any of the two types of sites), or when either $K = 0$, or $K = \infty$ (meaning that the one conformer is infinitely more stable than the other).

As we can see from Fig. 5.24, each curve starts from the value of one when $K = 0$, then rises as K increases, going through a maximum and eventually tends to one again for $K \to \infty$.

Figures 5.25 and 5.26 show the various pair correlation in this system. In Fig. 5.25 we show how the correlation $g(1, 1)$ depends on λ for different values of h (and $K = 1$). When $h = 1$, there is no correlation, g(1,1) = 1. When $h > 1$, the correlation becomes larger the larger h. This means that the more effective the ligand is in inducing conformational changes in the adsorbent molecule, the correlation

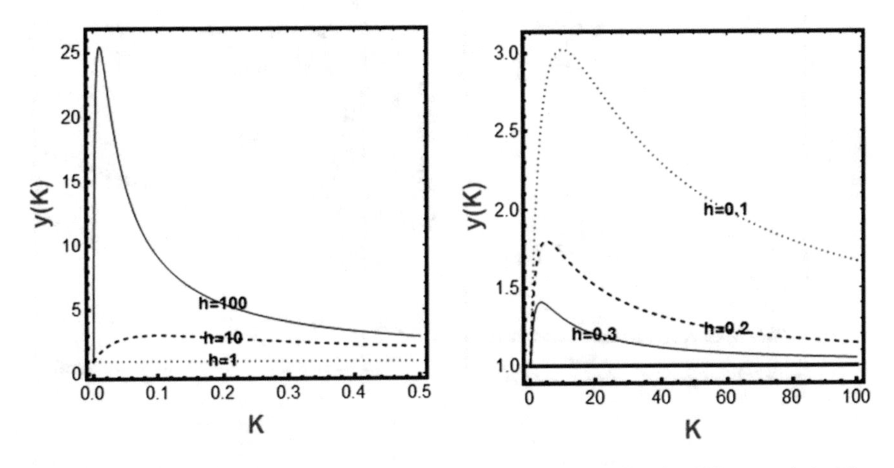

Fig. 5.24 The pair correlations defined in Eq. (5.39), as a function of K for different values of h

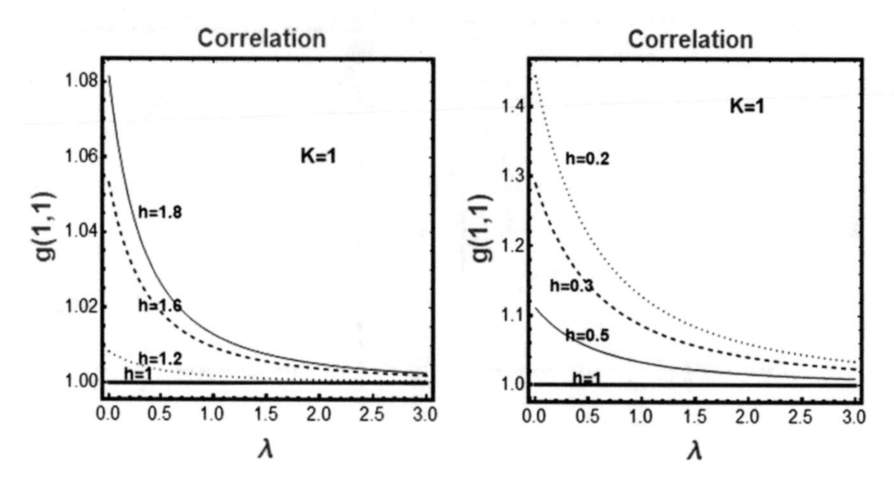

Fig. 5.25 The pair correlation g(1,1) as a function of λ for different values of h

$g(1, 1)$ is larger. Also, when $h < 1$, the smaller the h, the larger the correlation. The reason is similar; smaller value of $h < 1$ means more efficient in inducing conformational change. Note that the correlations $g(1, 1)$ are always positive (i.e. $g(1, 1) > 1$).

In Fig. 5.26 we show the two correlations $g(0, 0)$ and $g(1, 0) = g(0, 1)$, which are not usually studied in connection with cooperativity in binding phenomena. The general behavior of $g(0, 0)$ is similar to $g(1, 1)$. The larger $h > 1$, the larger the correlation, and the smaller $h < 1$, the larger the correlation.

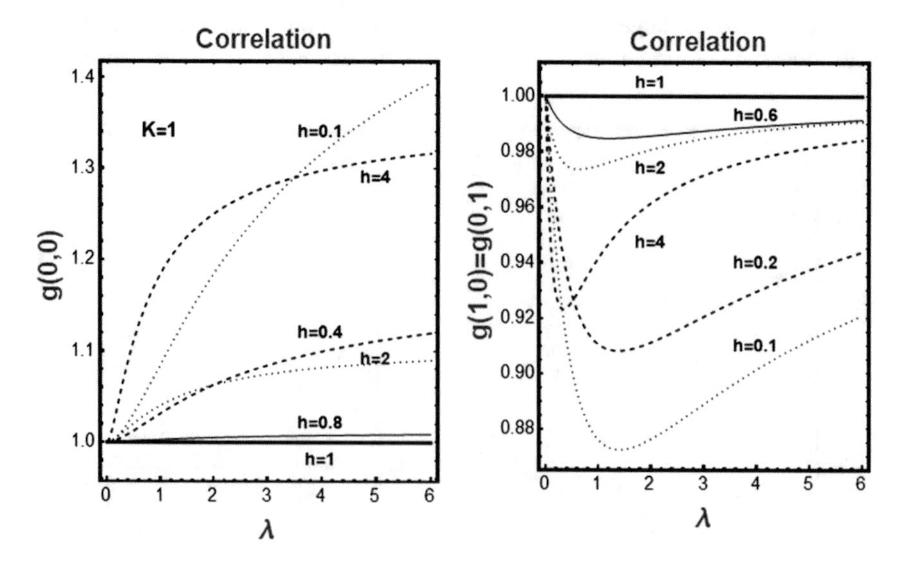

Fig. 5.26 The pair correlations, g(0,0) and g(1,0), as a function of λ for different values of h

The behavior of $g(1, 0)$ is quite different. Here, we see that all correlations are negative (i.e. $g(1, 0) = g(0, 1) < 1$). This means that given that site 1 is occupied, the probability of finding site 2 empty is smaller than the probability of finding site 2 empty, i.e.

$$g(1, 0) = \frac{Pr(\text{site 2 empty}|\text{site 1 occupied})}{Pr(\text{site 2 empty})} \tag{5.40}$$

This fact is important in the study of MI which is an average over $\log g(i, j)$, and which is always positive, see below.

Figure 5.27 shows the singlet SMI. As can be seen the curves are similar to those in Fig. 5.17. As we discussed in Sect. 3.5.1, all the curves start at zero for $\lambda = 0$. This means that one state (the empty state), has probability one. For $\lambda > 0$ the SMI rises to a maximum at SMI $= 1$, then tends to zero at $\lambda \to \infty$, meaning that the state "occupied" has probability one.

Figure 5.28 shows the pair SMI. The general behavior of these curves are the same as the SMI (1), except that the maximum reached is 2 instead of 1 in Fig. 5.27.

Finally, we show in Fig. 5.29 the mutual information $I(1; 2)$ as a function of λ for $K = 1$, and the various values of h. The shape of these curves are similar to those in Fig. 5.18. They all start at zero when $\lambda = 0$, and tend to zero at $\lambda \to \infty$. When $h = 1$ (meaning that the ligand has no preference to either L or H sites), we have no MI, $I(1; 2) = 0$. When there is a strong preference for either L or H, we get larger and larger MI.

Remember that this MI is due only to *indirect* communication between the sites. No direct interactions in this system.

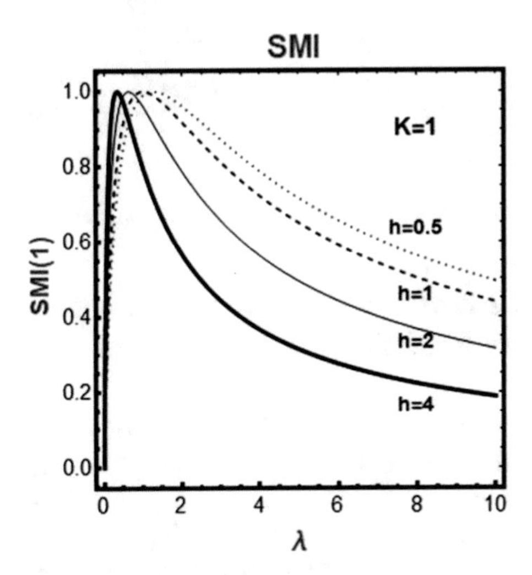

Fig. 5.27 The Singlet SMI as a function of λ for different values of h and $K = 1$

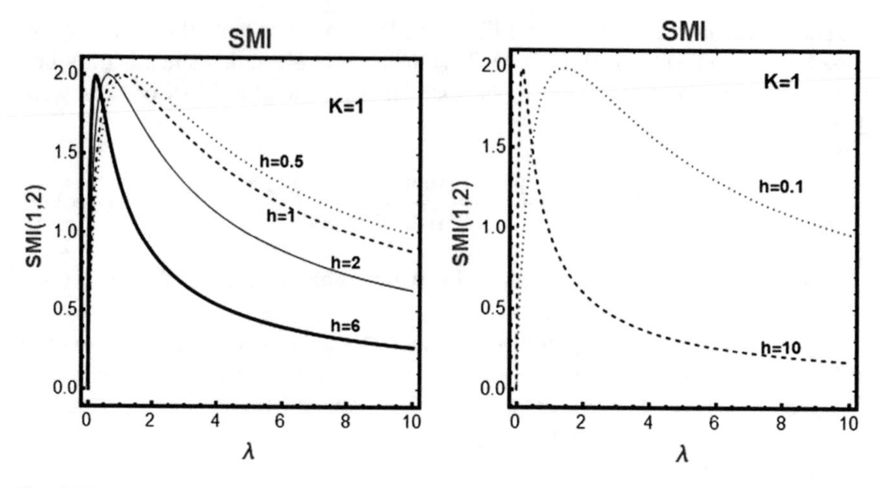

Fig. 5.28 The pair-SMI as a function of λ for different values of h

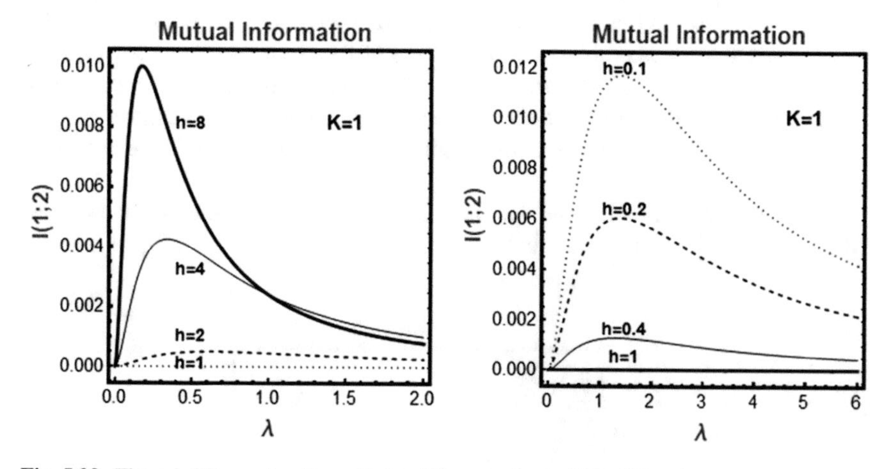

Fig. 5.29 The pair-MI as a function of λ for different values of h and K = 1

5.3.3 Indirect Communication Due to Conformational Changes in Each Subunit

In this section, we discuss a new type of communication between sites. The new model is similar to the one discussed in the previous section, also the results are very similar. However, the means of communication between the sites is different; this model is also the more relevant to biological binding systems.

The binding molecule consists of two subunits, each subunit has one binding site, and may be in one of two conformers.

As in the model of previous section, the ligand is assumed to bind with different energy to site L or to site H (by site L and H we mean site on the conformers L and H, respectively).

All the sixteen sites of the binding molecules were shown in Fig. 5.3. We use the same notations q_L and q_H for the binding strength to L and H, and Q_H and Q_L for the energetic states of H and L conformers. In addition, we have here three additional quantities which measure the interaction energy between two subunits. Thus, we define:

$$Q_{LL} = \exp\left[-\frac{E_{LL}}{k_B T}\right]$$
$$Q_{LH} = Q_{HL} = \exp\left[-\frac{E_{LH}}{k_B T}\right]$$
$$Q_{HH} = \exp\left[-\frac{E_{HH}}{k_B T}\right] \tag{5.41}$$

where $E_{\alpha\beta}$ is the interaction energy between a subunit is state α, and a subunit in state β (note that we assume that $E_{LH} = E_{HL}$, although one may think of a model where this assumption might not be correct, see Ben-Naim [1]). With the help of these three new quantities defined in Eq. (5.41), we define the following two parameters:

$$\overline{K} = \frac{Q_{HH}}{Q_{LL}}, \quad \eta = \frac{Q_{LH}^2}{Q_{HH}Q_{LL}} \tag{5.42}$$

The first measures the relative strength of the interactions E_{HH} and E_{LL}. The second is a measure of the equilibrium constant for the following reaction, written symbolically as:

$$(LL) + (HH) \leftrightarrows (LH) + (LH) \tag{5.43}$$

Denoting by $x_{\alpha\beta}$ the mole fraction of molecules in state $(\alpha\beta)$, the chemical equilibrium for the reaction in Eq. (5.43) is:

$$\eta = \frac{\left(x_{LH}^0\right)^2}{x_{LL}^0 x_{HH}^0} = \frac{(Q_L Q_H Q_{LH})^2}{Q_L^2 Q_{LL} Q_H^2 Q_{HH}} = \frac{Q_{LH}^2}{Q_{LL} Q_{HH}} \tag{5.44}$$

As we shall soon see, the new parameter η is important in determining the extent of communication between the two conformers, for details, see Ben-Naim [1].

As we had done in previous sections we extract all the probabilities from GPF which, is defined by:

$$\xi = Q(0) + Q(1)\lambda + Q(2)\lambda^2 \tag{5.45}$$

where

$$\begin{aligned} Q(0) &= \sum_{\alpha\beta} Q_\alpha Q_\beta Q_{\alpha\beta} \\ Q(1) &= \sum_{\alpha\beta} Q_\alpha Q_\beta Q_{\alpha\beta}(q_\alpha + q_\beta) \\ Q(2) &= \sum_{\alpha\beta} Q_\alpha Q_\beta Q_{\alpha\beta} q_\alpha q_\beta \end{aligned} \tag{5.46}$$

All the probabilities may be taken from the relevant terms in the GPF in the same way we have done in the previous sections.

Before discussing probabilities, correlation, SMI, and MI, it is helpful to discuss one important correlation that enters in the study of cooperativity. This is the correlation between two ligands occupying the two sites on the molecule:

$$g(1,1) = \frac{\Pr(1,1)}{\Pr(1)^2} \tag{5.47}$$

In general, this correlation dependents on the absolute activity λ. However, in the binding isotherm for this section, the quantity that determines the cooperativity in the system is the limit:

$$g^0(1,1) = \lim_{\lambda \to 0} g(1,1) = \frac{Q(1,1)Q(0,0)}{Q(0,1)Q(1,0)} \tag{5.48}$$

Using all the four parameters $K, h, \overline{K}$ and η, we can rewrite this correlation as:

$$g^0(1,1) = \frac{\left(1 + K^2\overline{K} + 2K\sqrt{\eta\overline{K}}\right)\left(1 + h^2K^2\overline{K} + 2hK\sqrt{\eta\overline{K}}\right)}{\left(1 + K\sqrt{\eta\overline{K}} + hK\left(K\overline{K} + \sqrt{\eta\overline{K}}\right)\right)^2} \tag{5.49}$$

Note that this is an *indirect* correlation. As in in Eq. (5.39), it is easy to see that there is no correlation when $h = 1$, or when $K = 0$ or $K = \infty$. Unlike the previous model, where the correlation was always positive (i.e. g^0 in Eq. (5.39) is larger than 1), here, the limiting correlation can either be positive or negative (i.e. $g^0 > 1$ or $g^0 < 1$). For details see Ben-Naim [1].

In Fig. 5.30a we show some selected values of $g^0(1,1) > 1$ and some values of $g^0(1,1) < 1$.

All the probabilities and the correlations, as well as the binding isotherm look much the same as in the previous section. We proceed to the SMI.

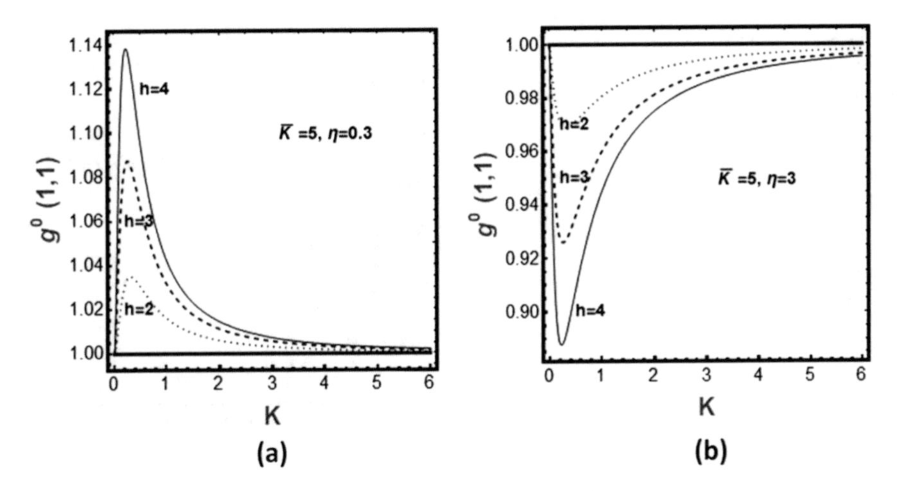

Fig. 5.30 Some limiting pair-correlations defined in Eq. (5.49), **a** positive correlations, and **b** negative correlations

Figure 5.31 shows the singlet and the pair SMI. As can be seen, the general behavior of both quantities is very similar to the previous case. Also, the SMI is similar to the previous case.

In Fig. 5.32 we see that the pair-MI is zero for both $\lambda = 0$ and $\lambda \to \infty$. When λ increases, the MI increases and then passes through a maximum before it decreases to zero when $\lambda \to \infty$. This is again very similar to the behavior of MI in the model discussed in the previous section.

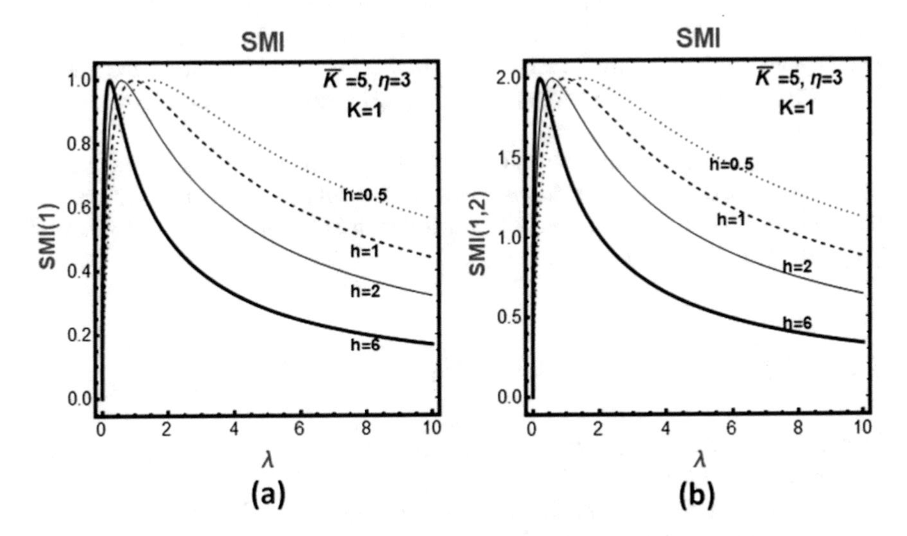

Fig. 5.31 The singlet (**a**) and the pair-SMI (**b**) as a function of $\lambda$for different values of h

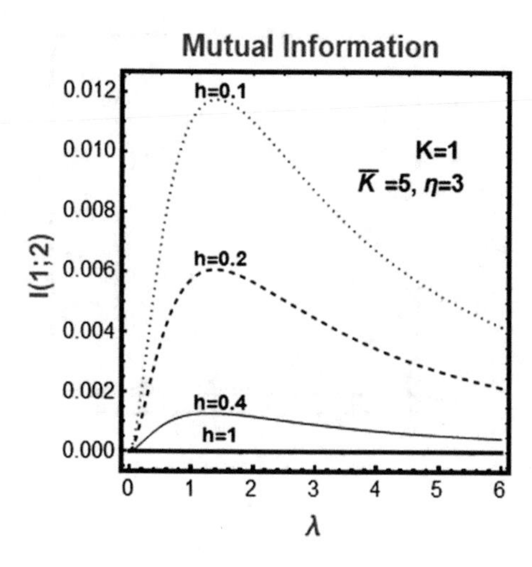

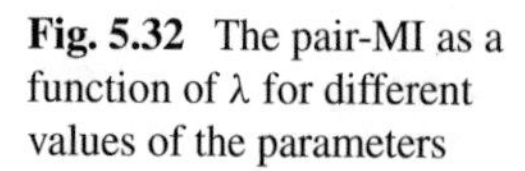
Fig. 5.32 The pair-MI as a function of λ for different values of the parameters

5.4 Three-Site Systems

In this section we discuss several systems with three binding sites. As in the previous section we start with the simplest case of a molecule with three identical sites, each can bind a ligand with some binding energy U. In addition, each pair of ligands bound to two sites interact by the parameter U_{int} which is the same for each pair. In the next variation of this model we assume that site 3 is different from either site 1 or site 2. It has a different binding energy to the site and in addition its presence affects the interaction between the ligands occupying sites 1 and 3. The next model will be the generalization of the model discussed in Sect. 5.3.3. This is the model of a molecule with three subunits, each subunit contains one binding site for a ligand. In this model we shall study the effect of the indirect interaction between the sites on the correlation between the various pair and triplet events such as "site i is occupied" or "site i and j are occupied", etc.

In the three-site system we shall also study the triplet SMI, the *total* mutual information (MI) and the *conditional* MI. In particular, we will be interested in the phenomenon of negative conditional MI which is believed to be a measure of "frustration" in the system [10, 11].

5.4.1 Direct Interaction Between Three Identical Sites

In this section, we discuss the simplest, three-site system with direct-interaction between pairs of ligands, Fig. 5.33.

The adsorbing molecule has three identical and equivalent sites. We use the same notation as in Sects. 5.1–5.3. The GPF is given by:

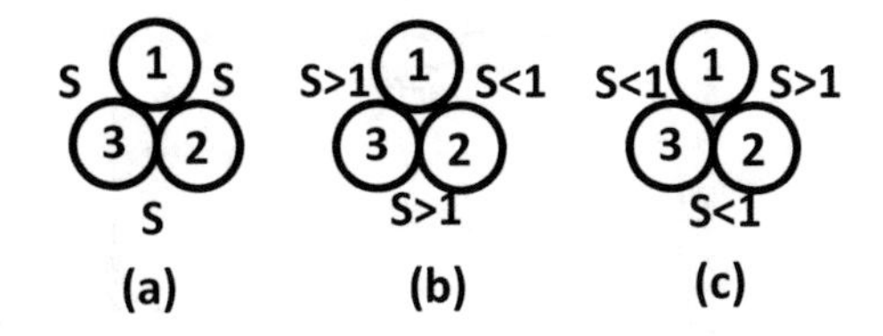

Fig. 5.33 The three different three-site-models with different direct pair-interactions

$$\xi = \sum_{i,j,j} Q(i,j,k)\lambda^{i+j+k} \tag{5.50}$$

where $Q(i, j, k)$ is the canonical PF of one molecule in state (i, j, k), where i, j and k may be zero or one, and the sites are denoted by 1, 2 and 3. λ is the absolute activity, which in this chapter may be viewed as a measure of the concentration of the ligand in either the gaseous or the liquid phase, which is at equilibrium with the ligands adsorbed on the molecule.

All the probabilities are derived from Eq. (5.50). For instance, the singlet probability $\Pr(1,\_,\_)$ is the probability of finding one specific site, say, 1 being occupied, regardless of the state of occupancy of the other sites. This is defined by:

$$\Pr(1,\_,\_) = Q(1,0,0)\lambda + [Q(1,1,0) + Q(1,0,1)]\lambda^2 + Q(1,1,1)\lambda^3 \tag{5.51}$$

and similarly:

$$\Pr(0,\_,\_) = Q(0,0,0) + [Q(0,1,0) + Q(0,0,1)]\lambda + Q(0,1,1)\lambda^2 \tag{5.52}$$

Note that in each of the Eqs. (5.51) and (5.52), site 1 is either occupied or empty respectively, and the sum is over all possible configurations of the other two sites.

Figure 5.34 shows these two probabilities as a function of λ for three values of S (defined by: $S = \exp\left[-\frac{U(1,2)}{k_B T}\right]$ where $U(1,2)$ is the ligand-ligand interaction), which measures the strength of the ligand-ligand interaction. $S > 1$ means attractive, and $S < 1$ repulsive interaction. The behavior of these probabilities is as expected, and similar to the corresponding probabilities in Fig. 5.23.

In Fig. 5.35 we show the binding isotherms for different values of λ. Again, the behavior is similar to that of Fig. 5.21.

Next, we show in Fig. 5.36 the various pair correlations in this system. The reader should note the difference in the behavior of the various correlations as a function of λ for different values of S. For instance, the pair correlation $g(0,0,\_)$ is larger than one (positive correlation) for $S > 1$, but smaller than one (negative correlation) when $S < 1$. A similar behavior is shown by $g(1,1,\_)$, but an opposite behavior of $g(1,0,\_)$, or $g(0,1,\_)$.

Figure 5.37 shows the triplet correlations in this system. The behavior is similar to the pair correlations. Note that there are both positive and negative correlation, these will play an important role when calculating the various mutual information in this system (see below).

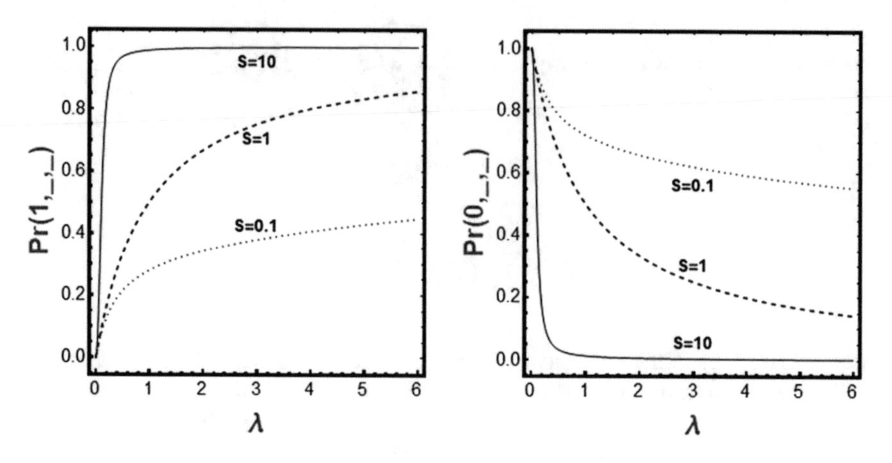

Fig. 5.34 The singlet probabilities as a function of λ for different values of S

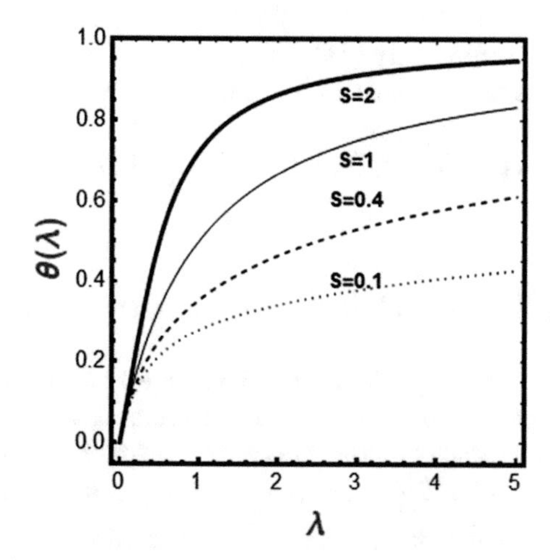

Fig. 5.35 The binding isotherm as a function of λ for different values of S

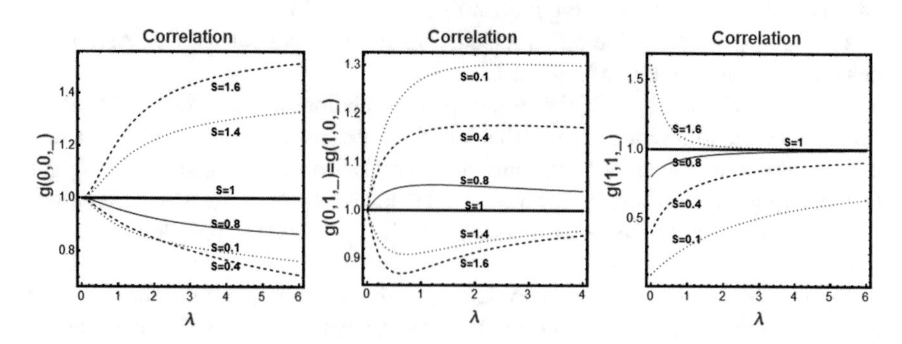

Fig. 5.36 The three different pair correlations as a function of λ, for different values of S

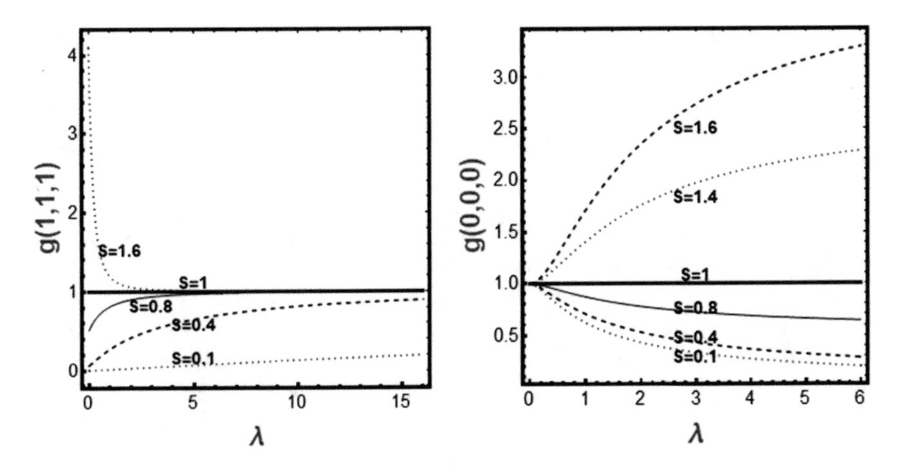

Fig. 5.37 The triplet correlations as a function of λ for different values of S

Figure 5.38 shows the three SMIs in this system. As expected all the curves start at zero (only one configuration—empty site—has probability one at $\lambda = 0$). Also, all the curves tend to zero when $\lambda = \infty$ (at this limit all the sites are occupied and the probability of the site being occupied is one).

In between $\lambda = 0$ and $\lambda = \infty$, the singlet SMI raises to one (meaning that there is a λ for which the probabilities of "empty" and "occupied" site becomes equal), then drops to zero as λ increases. Note that the drop is steeper the larger the value of S.

Figure 5.38b shows the pair SMI f or this system. The general behavior is similar to SMI(1). The only difference is in the maximum value which reaches 2 instead of 1 in Fig. 5.38a (due to four equally probable states or occupancy of the pair of sites). Figure 5.38c shows the triple SMI Again, the general forms of the curves are similar to those of SMI(1) and SMI(1, 2), except for the rise towards the value of nearly 3 (meaning eight equally probable state of occupancy of the three sites).

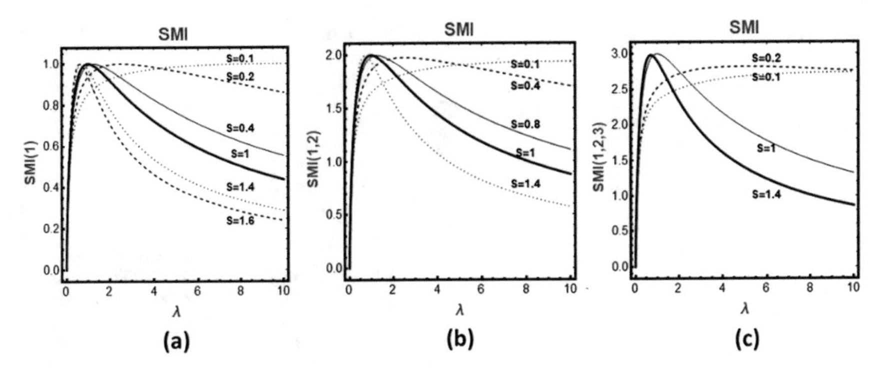

Fig. 5.38 The singlet (**a**), the pair (**b**) and the triplet SMI (**c**), as a function of λ, for different values of S

Figure 5.39 shows the pair-MI, $I(1;2)$ for this system. Remember the definition of $I(1;2)$ is:

$$I(1;2) = \sum_{i,j} p(i,j)\log g(i,j) \tag{5.53}$$

Each term in Eq. (5.53) may be either positive or negative (i.e. $g \geq 1$ or $g \leq 1$), but the MI must always be positive, Ben-Naim [10].

When $S = 1$, there is no correlation, and therefore $I(1;2)$ is zero. Once we turn-on the interaction we get a positive value of $I(1;2)$. The rise of $I(1;2)$ is steeper for $S > 1$, and much more shallow for $S < 1$. Note however (although not seen in this figure) that all the curves rise to a maximum value when λ increases, but in the limit $\lambda \to \infty$, all the curves tend to zero. At $\lambda \to \infty$ all sites are occupied with probability one, and the correlation becomes 1.

Figure 5.40 shows the *total* triplet MI, denoted $TI(1;2;3)$. This is defined as:

$$TI(1;2;3) = \sum p(i,j,k)\log g(i,j,k) \tag{5.54}$$

The TI are all equal to 0 when $S = 1$ (no correlation). All the curves start at the value of zero for $\lambda \to 0$, and end up at zero for $\lambda \to \infty$. In between, the curves reach a maximum value (it is not shown in Fig. 5.40b because the limit of 0 is reached at very high values of λ).

Perhaps, the most interesting quantity for this system is the *conditional* MI, denoted by $CI(1;2;3)$, shown Fig. 5.41.

The CI may be defined in three different, but equivalent ways (see Chap. 4 of Ben-Naim [10]). We shall use the definition in terms of *conditional pair* MI:

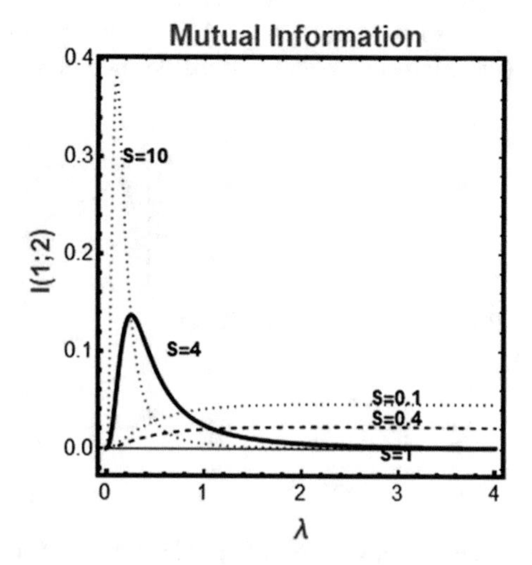

Fig. 5.39 The pair MI as a function of λ for different values of S

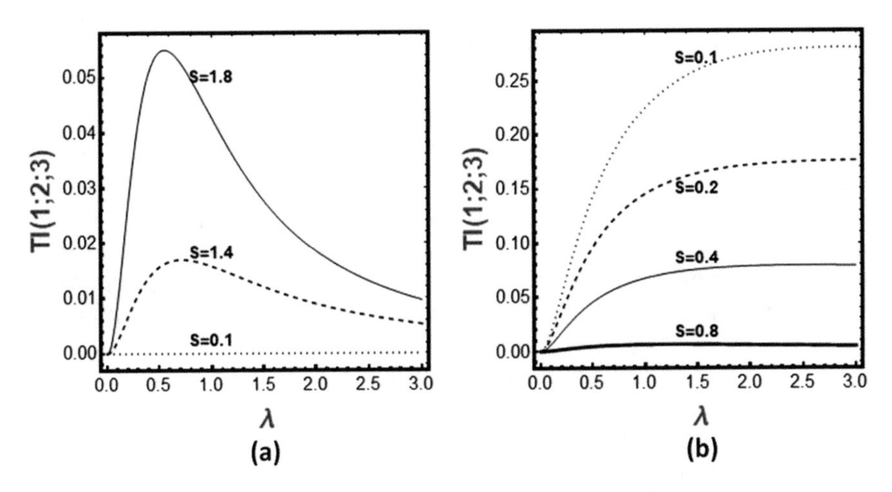

Fig. 5.40 The triplet total MI as a function of λ, for different values of S

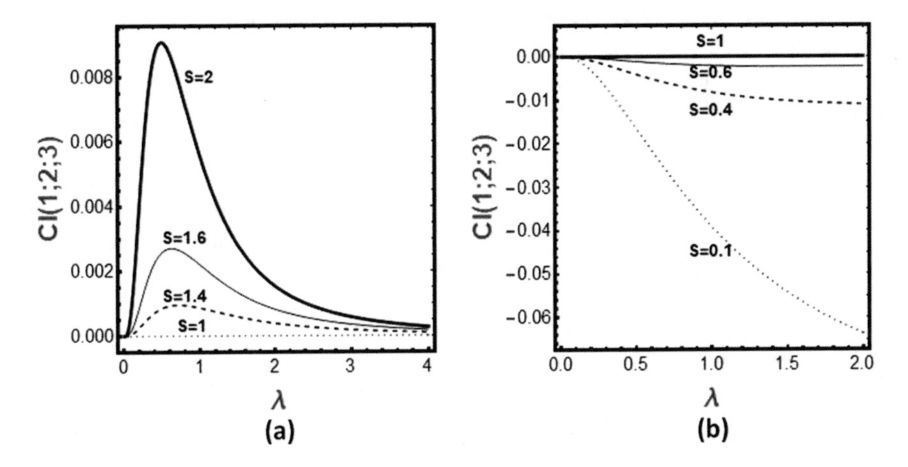

Fig. 5.41 The triplet conditional MI as a function of λ for different values of S

$$CI(1;2;3) = I(1;2) - I(1;2|3) \tag{5.55}$$

When $S = 1$, both $I(1;2)$ and $I(1;2|3)$ are equal to 1; there is no pair MI and no triplet MI. When $S > 1$, meaning there is attractive interaction between the ligands on the sites, the CI is positive. This means that "knowing the information on site 3, *decreases* the MI between 1 and 2. See Fig. 5.41a. When $S < 1$, meaning repulsive interaction, we see that knowing the information on site 3 *increases* the MI between 1 and 2, see Fig. 5.41b.

A negative value of CI was interpreted by Matsuda as a measure of frustration in the system. It is difficult to understand the behavior of this system in terms of frustration. We further discuss the suitability of this interpretation in Chaps. 1 and 3.

5.4.2 *Direct Pair and Triplet Interactions*

The model discussed in this section is similar to the one discussed in Sect. 5.4.1. Here, in addition to the pair-interaction which we denoted by S in Sect. 5.4.1, we add a triplet interaction denoted δ. When there are only two sites occupied, the direct interaction is S (as in Sect. 5.4.1). However, when the three sites are occupied, then there is an additional interaction energy. This addition energy affects only one term in the GPF (see Eq. 5.50). Instead of the factor S^3 for the fully occupied molecule we have the factor $S^3\delta$. In this case, the quantity Q(1,1,1), in Eq. (5.50) has the form:

$$Q(1,1,1) = Q(0,0,0)q^3S^3\delta \tag{5.56}$$

In the following illustrations we fix the value of S = 2, and examine the effect of the triplet interaction δ.

Figure 5.42 shows the pair MI, $I(1;2)$. Note that $\delta = 1$ means that no triple-interactions, i.e. the same model as in the previous section. We see that when $\delta = 10$, i.e. enhancing the attractive interaction, we get large values of MI. For $\delta = 0.1$, we get much smaller values of MI.

Figure 5.43 shows the two triplet MIs. First, the *total* MI in Fig. 5.43a, which is much the same as in the previous section. Note that the values of TI are for $\delta = 10$, everywhere larger than $\delta = 0.1$.

A completely different behavior is shown by conditional MI, $CI(1;2;3)$, Fig. 5.43b. Here, we see that for $\delta = 0.1$, we get *negative* values of CI. Can one interpret these results in terms of frustration? This result adds additional doubt to the interpretation of CI as a measure of frustration.

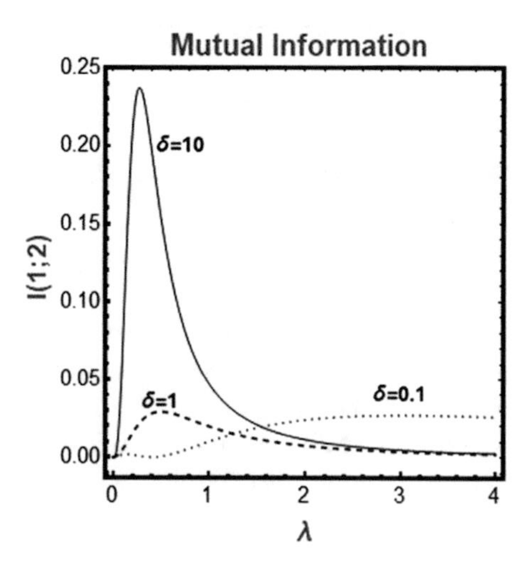

Fig. 5.42 The pair MI as a function of λ for different values of δ

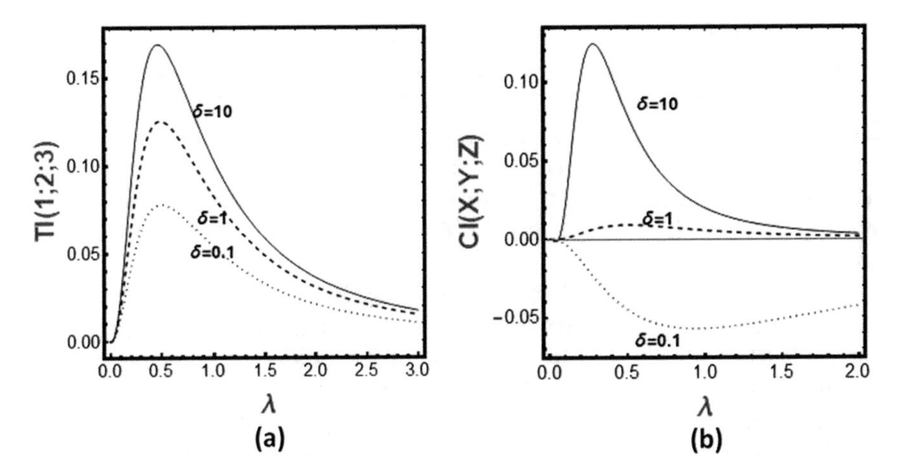

Fig. 5.43 The total (**a**) and the conditional (**b**) triplet MI as a function of λ, for different values of δ

5.4.3 Direct, But Different Pair Interactions

In this section, we study two variations of the model discussed in Sect. 5.4.1. Instead of the same parameter S, we had in the previous case for all the pairs, we assume here that there are two kinds of pair interactions; attractive and repulsive.

(i) ***One repulsive and two attractive***

The model is shown in Fig. 5.33b. We have the same three-site adsorbing molecule, but the interaction between the ligands are different. We have *attraction* between the pairs $(1, 3)$ and $(2, 3)$ with a corresponding factor $S > 1$, and the *repulsive* interaction between the pair $(1, 2)$ with a corresponding factor $\varepsilon < 0$. The GPF is the same as in Eq. (5.50) with the following two modifications:

$$
\begin{aligned}
Q(1,1,0) &= Q(0,0,0)q^2\varepsilon \\
Q(1,1,1) &= Q(0,0,0)q^3S^3\varepsilon
\end{aligned}
\tag{5.57}
$$

Normally, such a modification may be easily achieved for a system with two different ligands (say, one negatively charged and two positively charged ligands). However, we can also use the formalism of a one-type ligand by assuming that when a ligand is bound on site 3, it modifies the interaction between 1 and 2. This modification may be achieved either by polarization or my affecting the structure of the ligands at sites 1 and 2.

All the probabilities in this system behaves similarly to the previous models. Also, the SMIs are very similar [though there are slight differences between, say SMI$(1, 2)$ and SMI$(1, 3)$]. Therefore, we proceed directly to the MI.

Figure 5.44 shows the pair MI for the pairs (1, 2) and (1, 3) for $S = 2$, and various values of $\varepsilon \leq 1$. Note that all MIs are positive, as they should be. However, $I(1; 2)$ is quite different than $I(1; 3)$ (which is equal to $I(2, 3)$). The overall range of variation is also similar between 0 and 0.2. However, the dependence on ε is different. Making ε smaller (i.e. more repulsive interaction between 1, 2) *increases* the value of $I(1, 2)$. This means that the MI becomes larger, as the repulsive interaction increases. The opposite behavior is shown by $I(1, 3)$; the stronger the repulsive interaction the smaller the $I(1, 3)$.

Figure 5.45 shows the *total* and the *conditional* MI for this system. The general behavior is similar to the pair MI. All the curves of the total triplet MI start at $TI = 0$ for $\lambda = 0$, and tend to $TI = 0$, when $\lambda \to \infty$. All the curves pass through a maximum as a function of λ, and are positive as it should be.

A quite different behavior is shown by $CI(1, 2, 3)$, Fig. 5.45b. Here, we see that for all $\varepsilon < 1$, the conditional MI are *negative*. It is difficult to interpret these results in terms of frustration. One way of interpreting this result is based on the definition:

$$CI(1; 2; 3) = I(1; 2) - I(1; 2|3) < 0 \tag{5.58}$$

Since each pair MI must be positive, the negative value of CI means that knowing (3) *increases* the MI between 1 and 2. A possible physical situation for such a behavior is when ligand at site 3 causes polarization of the two ligands at 1 and 2, therefore, increases the correlation between 1 and 2.

One attractive and two repulsive

The model is shown in Fig. 5.33c. The interaction between (1, 2) is now attractive, say $S = 2$, but the interaction between both pairs (1, 3) and (2, 3) is repulsive.

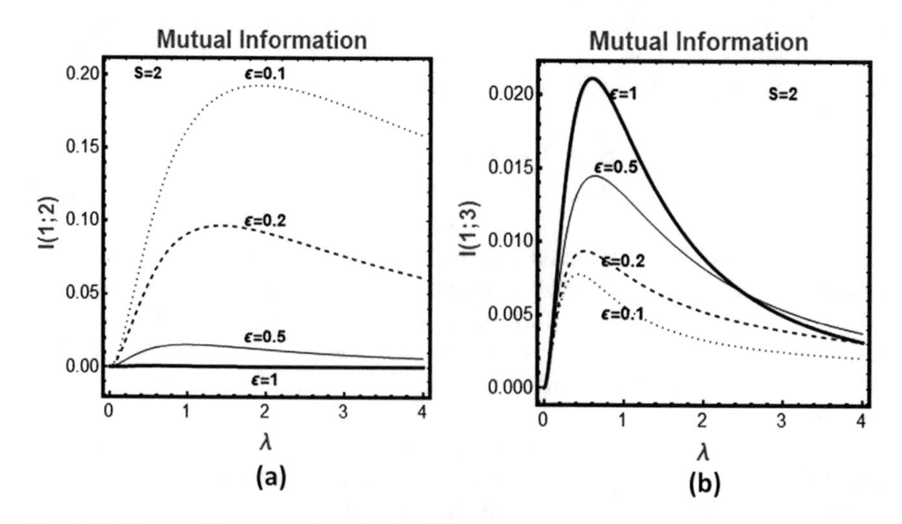

Fig. 5.44 The pair-MI as a function of λ for different values of ε

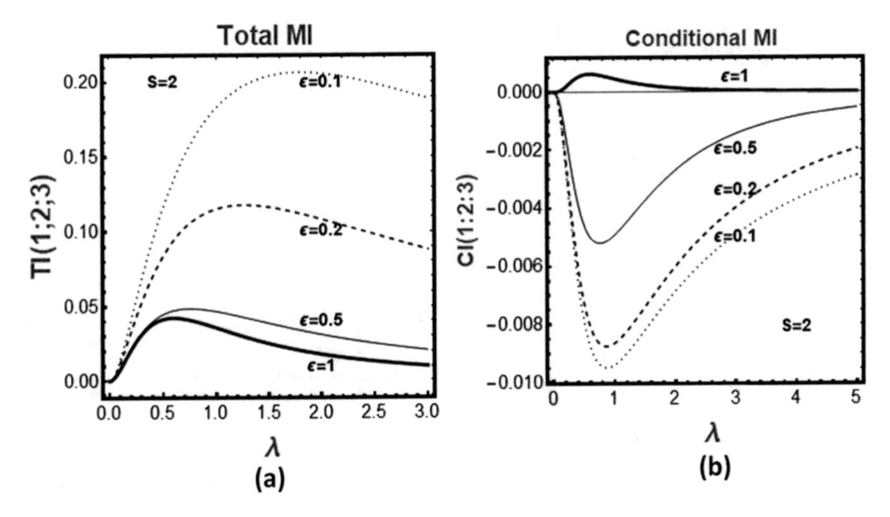

Fig. 5.45 The total and the conditional triplet MI as a function of λ, for different values of ε

The modification in the GPF is only in the following factors:

$$
\begin{aligned}
Q(1, 1, 0) &= Q(0, 0, 0)q^2 S \\
Q(1, 0, 1) &= Q(0, 1, 1) = Q(0, 0, 0)q^2 \varepsilon^2 \\
Q(1, 1, 1) &= Q(0, 0, 0)q^3 S \varepsilon^2
\end{aligned} \tag{5.59}
$$

Again, there is nothing different in the probabilities and the SMI. Therefore, we go directly to the various mutual information. As in the previous case, we have here two different pairs MI, $I(1; 2)$ and $I(1; 3) = I(2; 3)$.

Figure 5.46 shows the two pair MI. The general behavior of $I(1; 2)$ and $I(1; 3)$ is similar. All the curves start at 0, go through a maximum and then decrease to 0 at $\lambda \to \infty$. In both $I(1; 2)$ and $I(1; 3)$ we see that the magnitude of the MI decreases as ε increases. This means that *increasing* the repulsive interaction causes and increases the pair MI. This behavior is different from the one in the previous model, see Fig. 5.44.

Note also that for $I(1; 3)$, when $\varepsilon = 1$, there is no interaction between 1 and 3 (as well as between 2 and 3), and $I(1; 3)$ becomes 0. On the other hand, for $\varepsilon = 1$, $I(1; 2)$ is not zero because the pair (1, 2) is still interacting (attractively).

Compare this with Fig. 5.44 where it is $I(1; 2)$ which is zero for $\varepsilon = 1$, but not $I(1; 3)$.

Next, we show the total and the conditional triple MI in Fig. 5.47. The general behavior of TI is very similar to the previous case (Fig. 5.45a). Figure 5.47b shows the conditional MI for this case. Unlike the previous case where we observed *negative* CI for $\varepsilon < 1$, here, all the values of CI are *positive*. Can one say that in this case there is no frustration, whereas in the previous case there is?

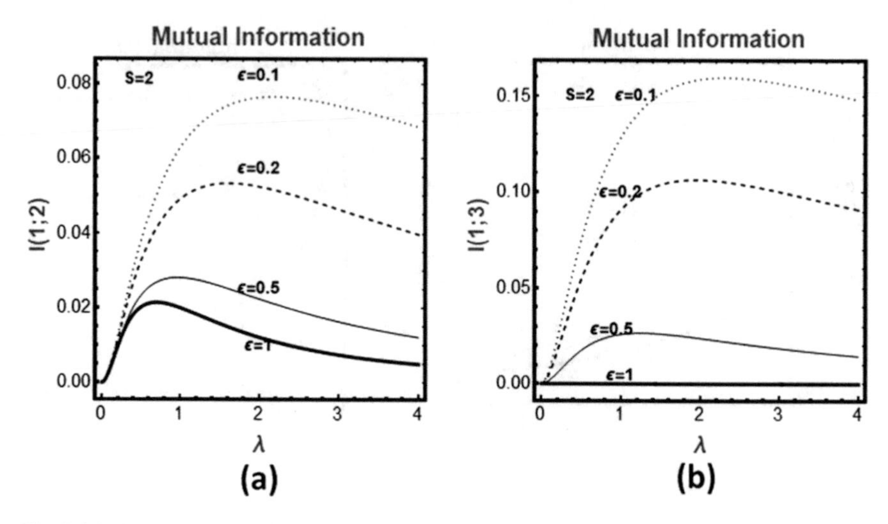

Fig. 5.46 The pair-MI as a function of λ, for different values of ε

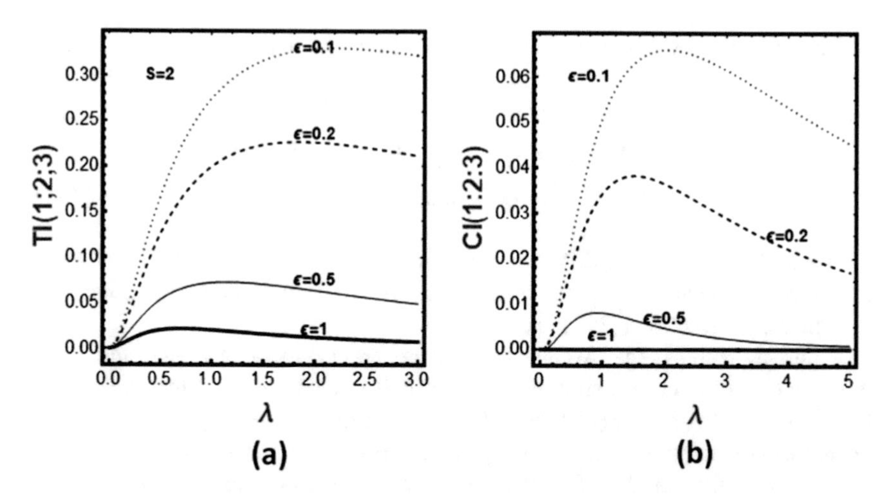

Fig. 5.47 The total and the conditional triplet MI as a function of λ, for different values of ε

5.5 Four-Site Systems with Direct Interactions

In this section we extend the study of the previous section to four-site systems. The study of these systems in terms of MI requires the extension of the concept of multivariate MI. In Ben-Naim [10], we defined two kinds of multivariate MI. A review of these two definitions is provided in Sect. 5.5.1 below. These are the *total* MI and *conditional* MI. The former is always a positive number, the latter is equivalent to Matsuda's definition [11]; it can be both positive and negative.

As in the previous section, we start with the GPF of a single adsorbing molecule with four binding sites which has the general form:

$$\xi = \sum_{i,j,k,m} Q(i, j, k, m)\lambda^{(i+j+k+m)} \tag{5.60}$$

Here, $Q(i, j, k, m)$ is the canonical PF of a single molecule having *configuration* (i, j, k, m), where each of these indices can either be a 0 or a 1. For instance, $Q(1, 0, 0, 0)$ means site 1 is occupied and all other sites are empty. $Q(1, 0, 1, 0)$ means sites 1 and 3 are occupied and sites 2 and 4 are empty. All the probabilities are obtained from the relevant terms in the GPF. The probability of finding the configuration (i, j, k, m) is:

$$P(i, j, k, m) = \frac{Q(i, j, k, m)\lambda^{(i+j+k+m)}}{\xi} \tag{5.61}$$

All the marginal probabilities are obtained from (5.61) by summation over all the unspecified variables, for instance the singlet probability $P(1, 0, 0, 0)$ is given by:

$$P(1, 0, 0, 0) = \frac{Q(1, 0, 0, 0)\lambda}{\xi} \tag{5.62}$$

This is the probability of finding a *specific* site (here site number 1) occupied and all other sites are empty. When we talk about the singlet probability we mean a specific site (say number 1) is occupied, and all the other sites are unspecified. This is defined as:

$$P(1, \_, \_, \_) = \sum_{j,k,m} P(1, j, k, m) \tag{5.63}$$

$P(1, j, k, m)$ means that site 1 is occupied and the configuration of all other sites is (j, k, m) and we sum over all possible values of j, k, m.

As we shall soon see, in some of the systems all sites are equivalent, in this case, the singlet probability is the same for each specific sites. For instance, in the case of a perfect square, studied in Sect. 5.5.2, all sites are equivalent, hence:

$$P(1, 0, 0, 0) = P(0, 1, 0, 0) = P(0, 0, 1, 0) = P(0, 0, 0, 1) \tag{5.64}$$

And also:

$$P(1, \_, \_, \_) = P(\_, 1, \_, \_) = P(\_, \_, 1, \_) = P(\_, \_, \_, 1) \tag{5.65}$$

However, even when all sites are equivalent (like the perfect square) the pair probabilities are not necessarily all equal. For instance, the pair probabilities of nearest neighbors (nn) sites, are equal, say: $P(1,1,0,0) = P(0,1,1,0)$. But the pair probability of next-nearest-neighbor (nnn) is, in general, different, e.g. $P(1,0,1,0) \neq P(1,1,0,0)$.

5.5.1 Multivariate Mutual Information (MI)

In this section, we review the various definitions of the multivariate MI for the case of four-site system. There are several different definitions of multivariate MI. We shall use here only two of these.

1. **The *total* mutual information**

   For two *random variables* we define:

$$TI(1;2) = I(1;2) = \sum_{x_1,x_2} p(x_1,x_2) \log g(x_1,x_2)$$
$$= H(1) + H(2) - H(1,2) \tag{5.66}$$

   The extension for three *random variables* we have:

$$TI(1;2;3) = \sum_{x_1,x_2,x_3} p(x_1,x_2,x_3) \log g(x_1,x_2,x_3) \tag{5.67}$$

   And for *n random variables* we have:

$$TI(1;2;\ldots n) = \sum_{x_1,\ldots x_n} p(x_1,\ldots x_n) \log g(x_1,\ldots x_n) \tag{5.68}$$

2. **The *conditional* mutual information**

   We start from the identity, see Sect. 4.2 of Ben-Naim [10]:

$$I(1;2) = I(1;1) - I(1;1|2) \tag{5.69}$$

where $I(1;1)$ is the "self-information" which is defined as:

$$I(1;1) = H(1) + H(1) - H(1,1) = H(1) \tag{5.70}$$

   We now generalize for three *random variables*:

$$CI(1;2;3) = I(1;2) - H(1;2|3) \tag{5.71}$$

   We refer to this as *conditional* MI. It measures the effect of knowing X_3, on the MI between X_1 and X_2. It is easy to show that CI may be expanded in terms of SMI (see Eq. (4.15) of Ben-Naim [10]). The result is:

$$\begin{aligned} CI(1;2;3) = {} & H(1) + H(2) + H(3) - H(1,2) \\ & - H(1,3) - H(2,3) + H(1,2,3) \end{aligned} \quad (5.72)$$

The last equation is reminiscent of the inclusion–exclusion principle in probability. This result had led some authors to use Venn diagram for SMI and MI, which, as we discussed in Chap. 1, is not warranted.

For three *random variables* we generalized the CI as:

$$C(1;2;3;4) = CI(1;2;3) - CI(1;2;3|4) \quad (5.73)$$

and for any *n random variables*, we define:

$$CI(1;2;\ldots n) = CI(1;2;\ldots;n-1) - CI(1;2;\ldots;n-1|n) \quad (5.74)$$

It is easy to show that this definition is equivalent to Matsuda's definition in terms of the inclusion–exclusion principle. We show here the equivalency for three *random variables*. Starting from the definition (5.73) and using (5.71), we can rewrite (5.73) as:

$$C(1;2;3;4) = I(1;2) - I(1;2|3) - [I(1;2|4) - I(1;2|3,4)] \quad (5.75)$$

In (5.75), we have on the right-hand-side only pair MI. These can be written in terms of MI by expansion, similar to Eq. (5.72), to obtain:

$$\begin{aligned} C(1;2;3;4) = {} & H(1) + H(2) - H(1,2) - [H(1|3) + H(2|3) \\ & - H(1,2|3)] - [H(1|4) + H(2|4) - H(1,2|4)] \\ & + [H(1|3,4) + H(2|3,4) - H(1,2|3,4)] \\ = {} & H(1) + H(2) + H(3) + H(4) - H(1,2) - H(1,3) - H(1,4) \\ & - H(2,3) - H(2,4) - H(3,4) + H(1,2,3) + H(1,2,4) \\ & + H(1,3,4) + H(2,3,4) - H(1,2,3,4) \end{aligned} \quad (5.76)$$

This is the Matsuda definition of the triplet MI. We prefer to call this a conditional MI because of the relatively simpler interpretation of the definition in terms of conditional MI. this form has led some people to use Venn diagrams for the MI, see Sect. 1.4.

5.5.2 A Perfect Square with Equal Nearest Neighbor Interactions

We start with the simplest four-site system arranged in a perfect square, Fig. 5.48. All the pair interactions are the same. We assume only direct interaction between pairs of nn ligands.

We start with the singlet probabilities. Figure 5.49 shows the two probabilities $P(0, \_, \_, \_)$ and $P(1, \_, \_, \_)$. The general dependence on λ is similar to what we have observed in the two-site and three-site systems.

The pair probabilities are shown in Fig. 5.50. Here, we see that there are differences between the nn-pair probability and nnn-probability. However, the general dependence of these probabilities on λ is similar. The difference between these probabilities leads to the difference in the corresponding correlation functions, and these in turn affect the pair-MI which we are discussing below.

There are numerous, and different pair correlation functions in this system. In Fig. 5.51, we show only two of these; the nn-pair correlation and the nnn-pair correlation.

Note that for the nn-pair correlation we have both positive and negative correlations. When $S < 1$, the probability of finding two sites occupied is smaller than the

Fig. 5.48 The four binding sites arranged in a perfect square

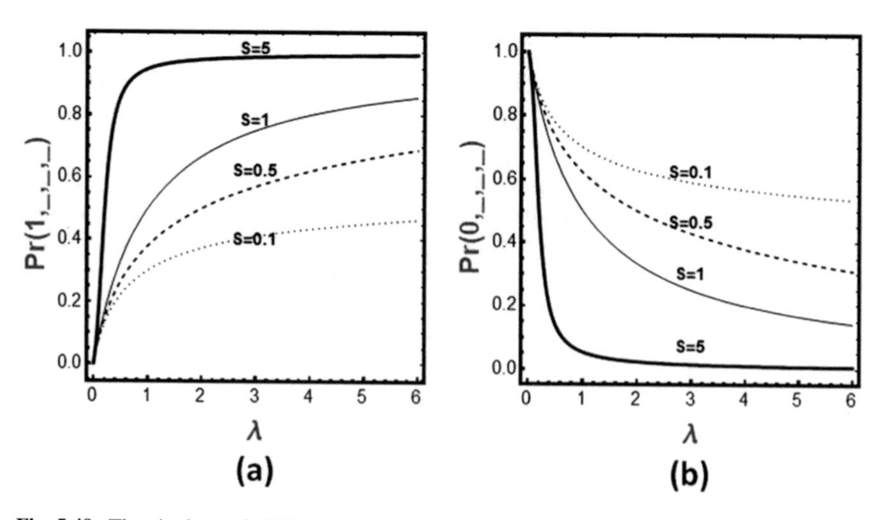

Fig. 5.49 The singlet probabilities for the square model as a function of λ, for different values of S

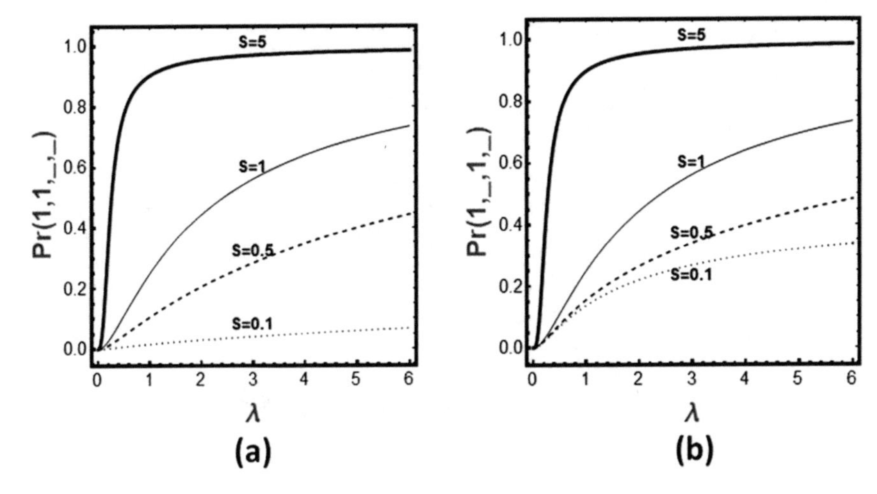

Fig. 5.50 The two pair probabilities; **a** the nn, and **b** the nnn, for the square model as a function of λ for different values of S

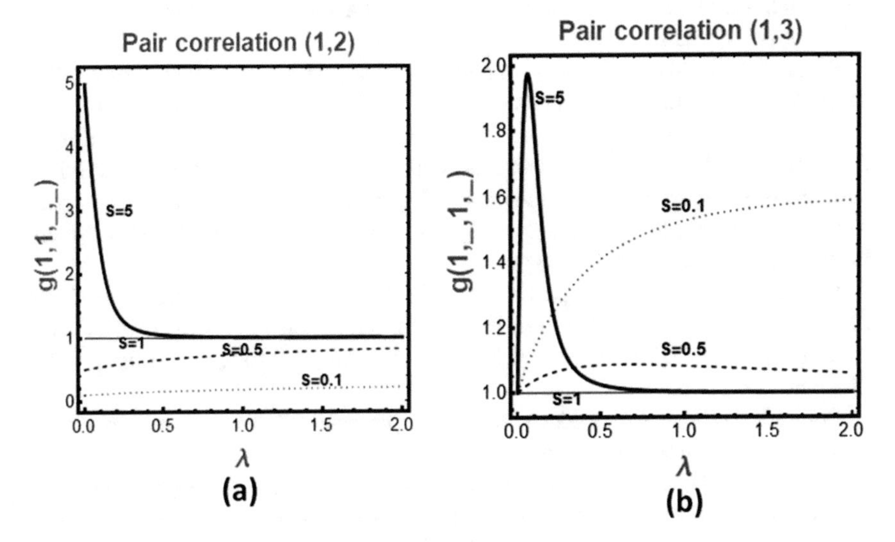

Fig. 5.51 The two pair correlations: **a** the nn and **b** the nnn, for the square model as a function of λ, for different values of S

product of the two singlet probabilities (i.e. $g < 1$). For $S < 1$, we have positive correlation (i.e. $g > 1$) and for $S = 1$ there is no correlation.

The nnn-pair correlation are all positive for either $S < 1$, or $S > 1$, the pair correlation is larger than one. Note also that in the limit of $\lambda \to 0$, the nn-pair correlation tends to S, whereas the nnn-pair correlation tends to 1, i.e. no correlation.

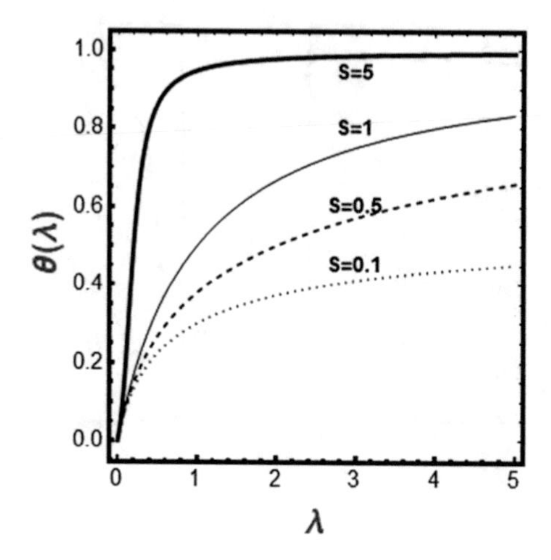

Fig. 5.52 The binding isotherm for the square model, for different values of S

On the other limit of $\lambda \to \infty$, all pair correlations tend to one (no correlation). This behavior is important.

Figure 5.52 shows the binding isotherms for this system. The general form of the curve is similar to the previous models.

Figures 5.53 and 5.54 show the SMIs for this system. The general form of all the SMI is similar to the previous models. They are also very similar for different numbers of sites, except for the maximal values of 1, 2, 3 and 4 for the singlet, pair, triplet, and quadruplet SMI, respectively. (Note that SMI(1, 2) is slightly different from SMI(1, 3), not shown here).

Figure 5.55 shows the pair-MI, $I(1;2)$ and $I(1;3)$, and the conditional $MII(1;2|3)$. Again, the general form of all the curves are quite similar. All curves start at 0 for $\lambda = 0$, and tend to 0 for $\lambda \to \infty$. All go through a maximum as a function of λ.

Figure 5.56 shows the triplet and the quadruplet total MI. The general form of the curves is similar in the two figures.

Figure 5.57 shows the conditional MI. Here, all the values of the CI are *positive*, but we note that the difference $CI(1;2;3) - CI(1;2;3;4)$ is everywhere positive.

It should be noted that while the *total* MI must always be positive (see Chap. 4, Ben-Naim [10]), the *conditional* MI might either be positive or negative. Examples of negative CI will be shown in the next sections.

5.5.3 A Perfect Square with Unequal Pair-Interactions

In the previous model we assumed that all the interactions between pairs of ligands on nn-sites are equal (either all repulsive or all attractive). In the present model we

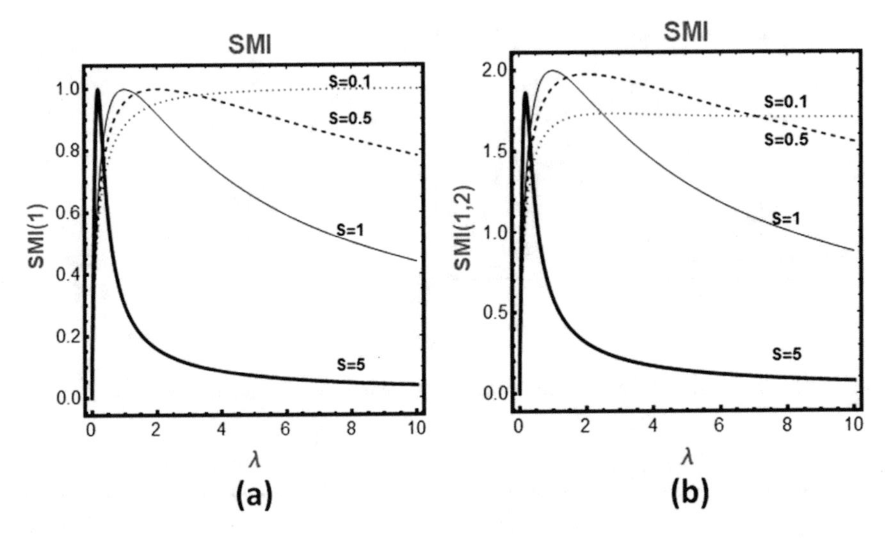

Fig. 5.53 The singlet (**a**) and the pair SMI (**b**), as a function of λ for different values of S

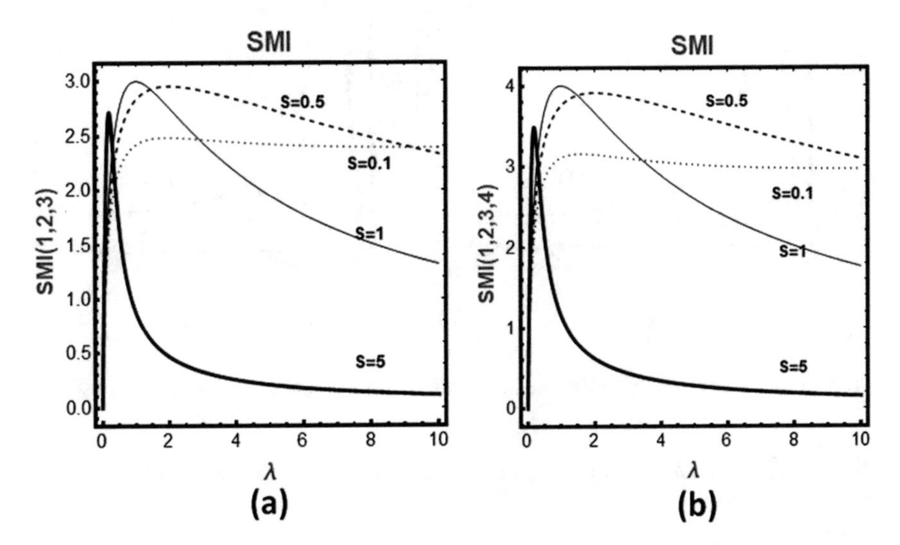

Fig. 5.54 The triplet (**a**) and the quadruplet SMI (**b**), as a function of λ, for different values of S

assume that one pair (say, 1,2) interacts repulsively, whereas all other pairs interact attractively. As we have noted in Sect. 5.4.3, such a scheme of interaction can easily be achieved with two different ligands. This will require changing the theory from which we derive the probabilities. However, we use here the same formalism we used in the previous sections when we have one type of ligands, and instead assume that the structure of the adsorbent molecule is different. For instance, if the sites 1 and 2 are very near to each other, then two ligands on these two sites will repel each

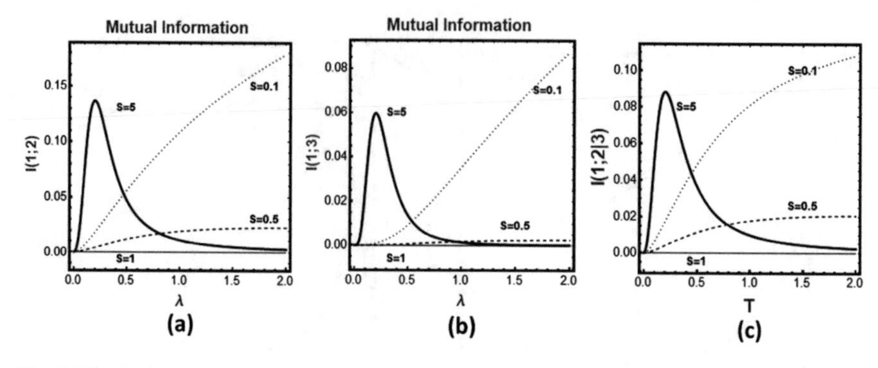

Fig. 5.55 The pair-MI and the conditional MI as a function of λ, for different values of S

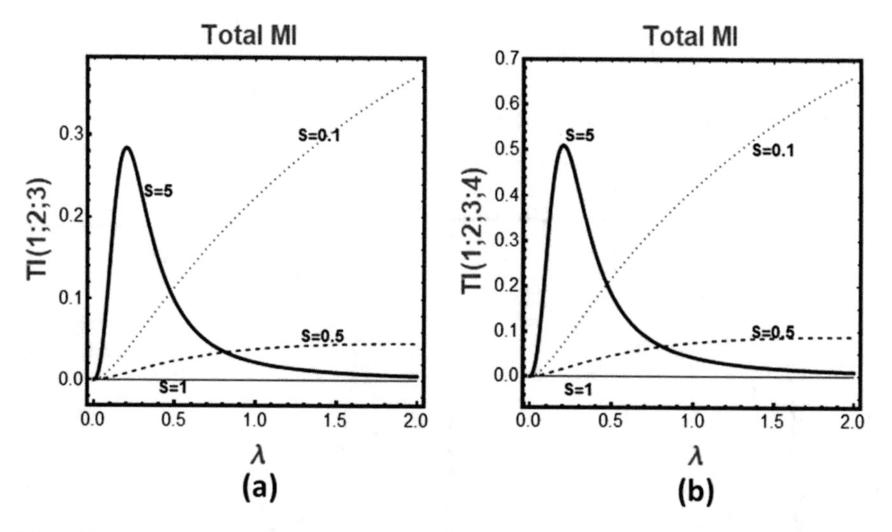

Fig. 5.56 The total triplet (**a**) and the quadruplet MI (**b**), as a function of λ, for different values of S

other. Binding on any other pair of sites which are at larger distance will produce an attractive interaction.

(i) **One pair of repulsive, and three pairs of attractive interaction**

For the following computation we fix the interaction parameter for the pair (1, 2) at $S = 0.1$, and choose three values of interaction parameters of all the other pairs, Fig. 5.58. Clearly, in this system the sites 1 and 2 are equivalent, but they are different from sites 3 and 4.

All the probabilities in this system are similar to the previous case. Therefore, proceed directly to the SMIs. Figure 5.59 shows the SMI (1) (equal to SMI (2)), and SMI (3) (equal to SMI (4)). The general form of the curves is similar to the case of the previous case. This is a result of the similarity of the singlet probabilities.

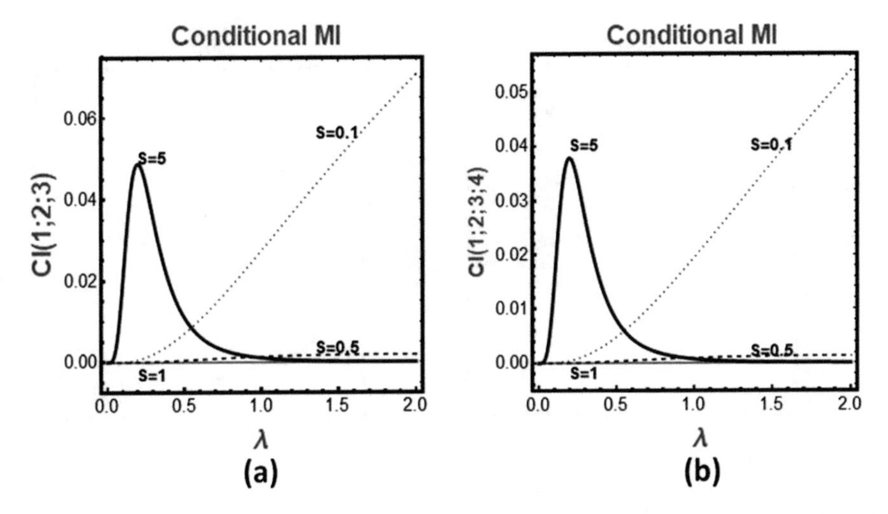

Fig. 5.57 The conditional triplet (**a**), and the quadruplet MI (**b**), as a function of λ, for different values of S.

Fig. 5.58 The Four sites arranged with different pair interactions

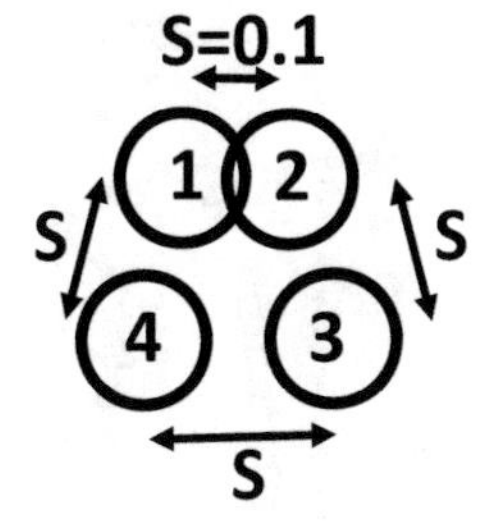

The pair, triplet, and quadruplets are all similar in form to the singlet SMI. They all start at 0 for $\lambda \to \infty$. In between they go through a maximum.

Figure 5.60 shows two of the pair MI. Note that the values of $I(2:3)$ are systemically smaller than the value of $I(1;2)$.

In Fig. 5.61, we show the total triplet and quadruplet MI. Here, as expected all the values are positive. Figure 5.62 shows the conditional MI for three and four sites. The new feature here are the negative values of CI. The triplet MI shows a minimum at lower values of λ, and then a maximum (positive) at higher values of λ, then tends to 0 for $\lambda \to \infty$. The behavior of the conditional quadruplet MI is different. We first observe a minimum (negative) at lower λ, but then they tend to 0 at higher λ.

It should be noted that Matsuda [11] interprets a negative value of CI in terms of frustration (see Chap. 3). Here, it is difficult to claim that there is frustration similar to the case in spin system. If we interpret negative values of CI with repulsive interaction then it is not clear why the values of CI for the three sites (1, 2, 3) have both positive and negative values.

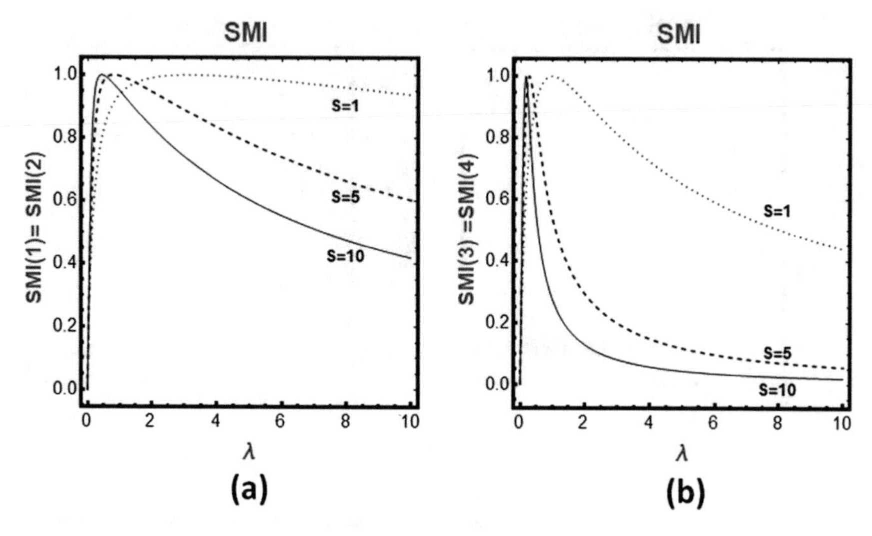

Fig. 5.59 The two different singlet SMI as a function of λ, for different values of S

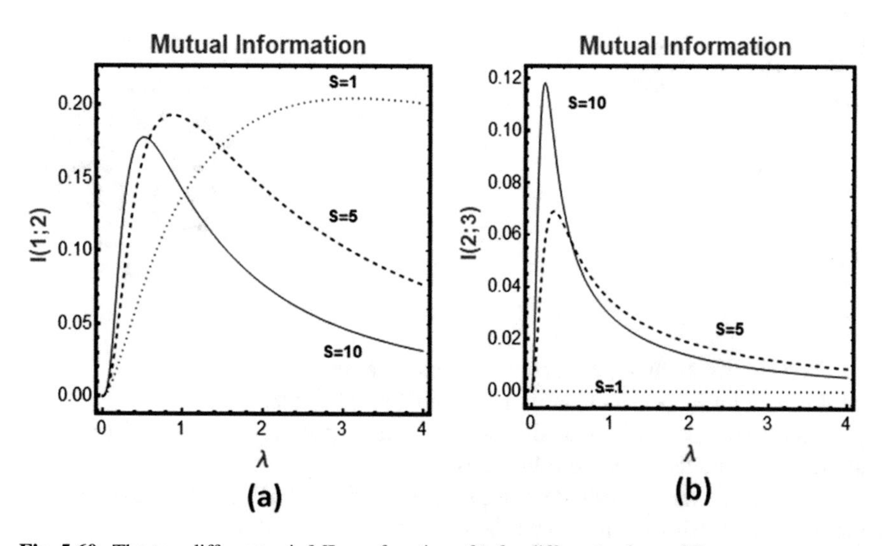

Fig. 5.60 The two different pair-MI as a function of λ for different values of S

(ii) **Two pairs with repulsive: (1,2) and (2,3), and the other two attractive interaction**

Next, we examine the case of two pairs, (1,2) and (2,3), with repulsive interactions with a parameter $S = 0.1$, and the two other pairs interact with varying attractive interaction $S = 1, 5, 10$, Fig. 5.63.

All the curves of the SMI are similar to the previous case. The only new behavior is of the conditional MI, Fig. 5.64.

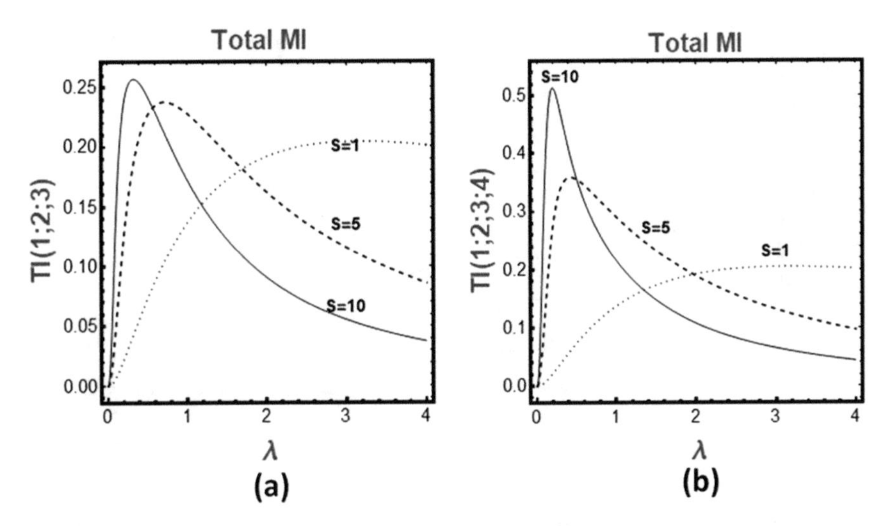

Fig. 5.61 The total triplet MI (**a**), and the quadruplet MI (**b**), as a function of λ, for different values of S

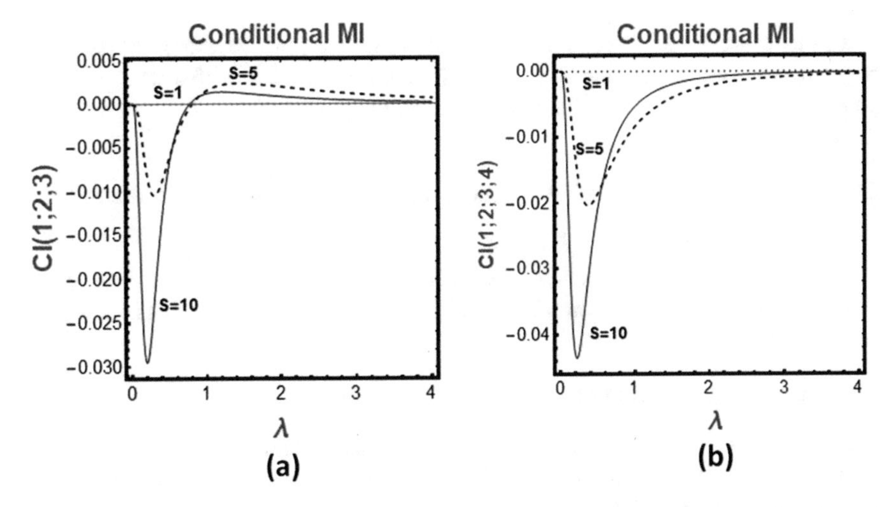

Fig. 5.62 The conditional triplet MI (**a**), and the quadruplet MI (**b**) as a function of λ, for different values of *S*

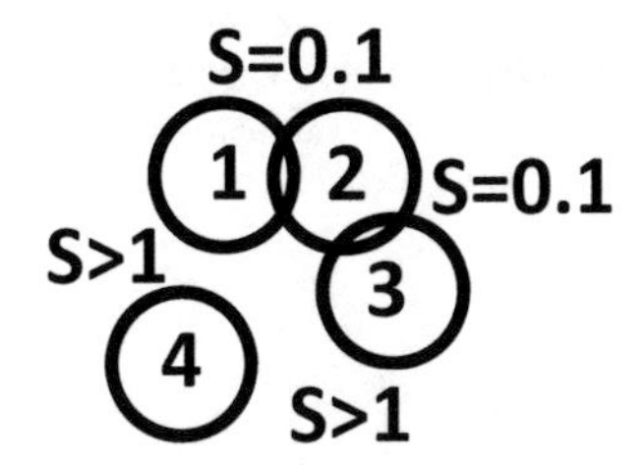

Fig. 5.63 The Four sites arranged with different pair interactions; two attractive and two repulsive

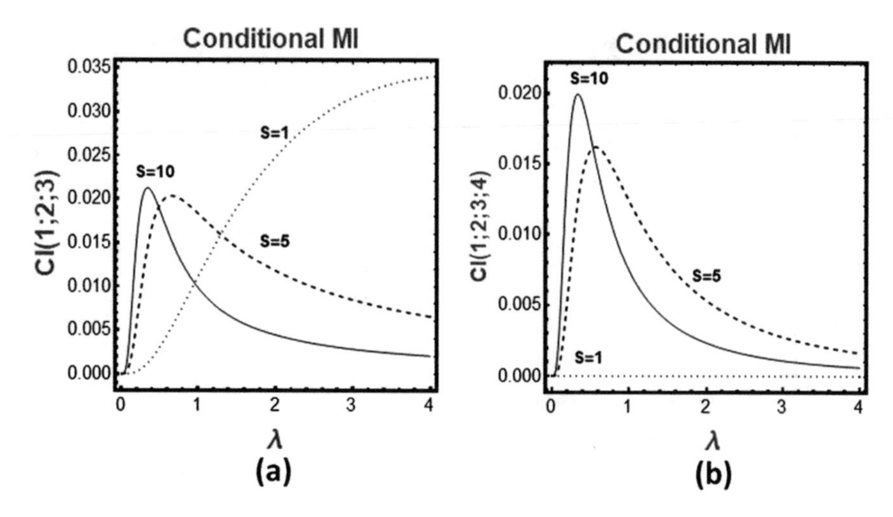

Fig. 5.64 The conditional triplet MI (**a**), and the quadruplet MI (**b**), as a function of λ, for different values of S, for the case of Fig. 5.63

In this case, we see that all the values of the CI are positive. Here, we have two negative interactions, but the CI is everywhere positive. It seems that when we have an even number of repulsive interaction we get the positive CI. This is confirmed in the following case.

(iii) Two pairs with repulsive interaction (1, 2) and (3, 4) and two attractive

The arrangement is shown in Fig. 5.65.

In Fig. 5.66 we show the conditional MI for this case. Again, we observe all positive values. Perhaps, confirming the conclusion that for even number of repulsive interaction the CI is positive.

(iv) Three repulsive and one attractive interaction

The final model has three repulsive and one attractive interaction, Fig. 5.67.

The behavior of the conditional MI is quite different here. Figure 5.68a shows that for $S = 1$ we have positive values of conditional triplet CI, but for $S = 5$ and $S = 10$,

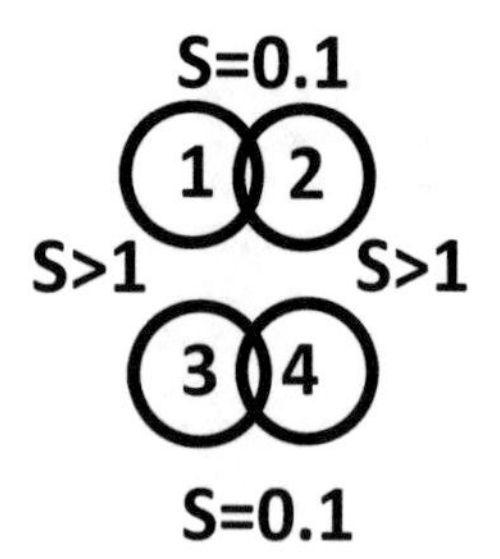

Fig. 5.65 The Four sites arranged with different pair interactions; two attractive and two repulsive

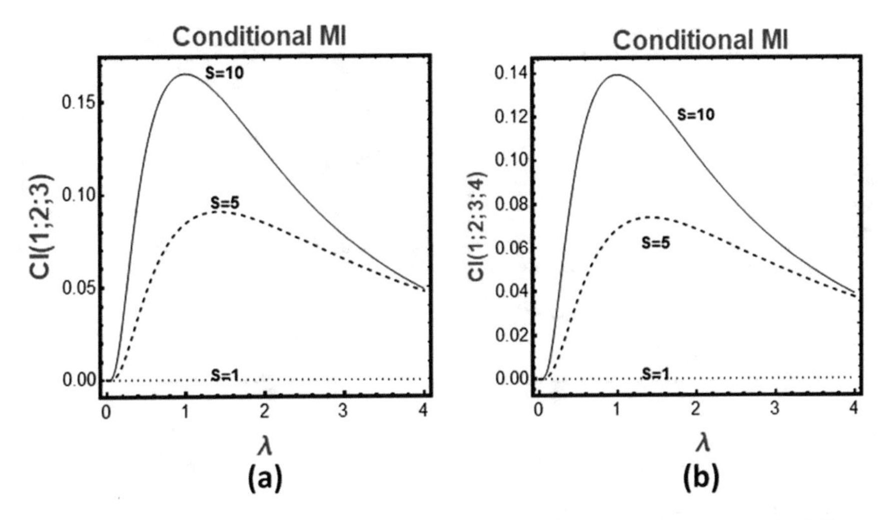

Fig. 5.66 The conditional triplet MI (**a**), and the quadruplet MI (**b**), as a function of λ, for different values of S, for the case of Fig. 5.65

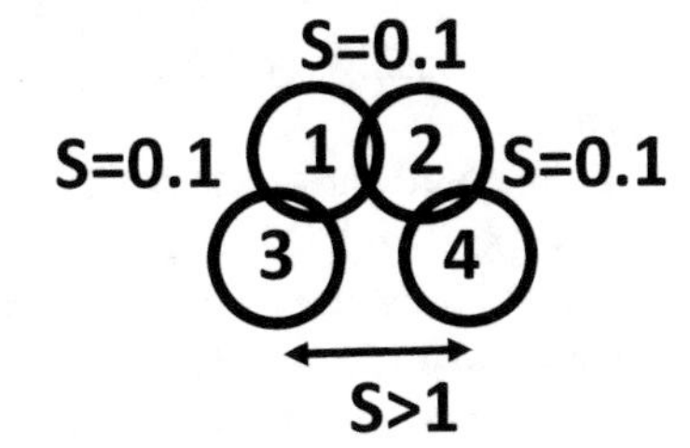

Fig. 5.67 The Four sites arranged with different pair interactions; three repulsive and one attractive

we have negative values of CI. In Fig. 5.68b we see both positive and negative values of conditional quadruplet CI. This it is difficult to conclude anything regarding the relation between negative values of the CI and either frustration or either odd or even number of repulsive interactions.

5.5.4 Parallelogram with Five Equal Pair Interactions

In this section, we consider the case of a square with one additional interaction between the pair 1, 3. This is equivalent to a parallelogram for which the distance between 1, 3 is the same as the length of all the edges. All the probabilities, the SMI, and the pair MI behave similarly to the case of the perfect square with four mm interactions. Therefore, we present here some results on the triplet and quadruplet MI.

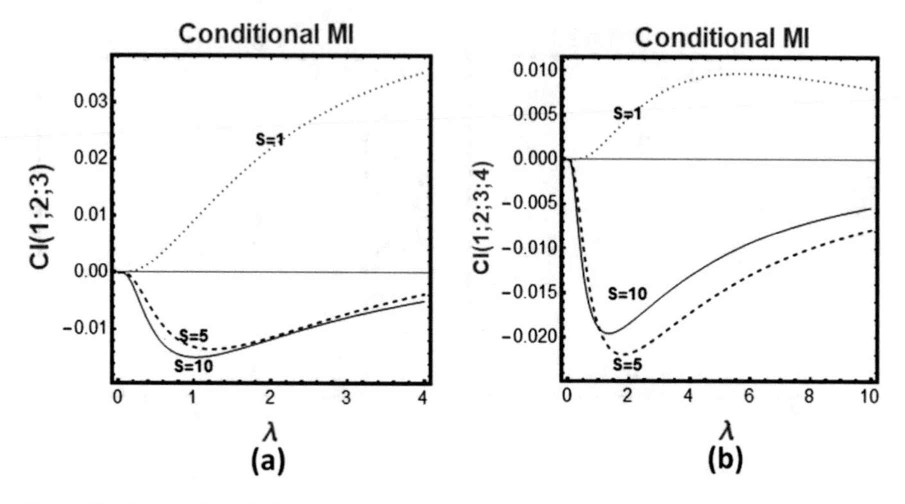

Fig. 5.68 The conditional triplet MI (**a**), and the quadruplet MI (**b**) as a function of λ, for different values of S, for the case of Fig. 5.67

Figure 5.69 shows the total triplet and the quadruplet total MI. As expected, all values are positive and the general forms of the curves are similar to the case of a square.

Figure 5.70 shows the triplet conditional MI. Here, unlike the perfect square we find the positive $CI(1, 2, 3)$ for attractive interaction but, negative CI for the case of repulsive interaction. A quite different behavior is shown by $CI(2; 3; 4)$, which are everywhere positive. Clearly, the triplet (1, 2, 3) is different from the triplet (2, 3, 4). It is also clear that the former has three interactions [between the pairs (1, 2), (2, 3)

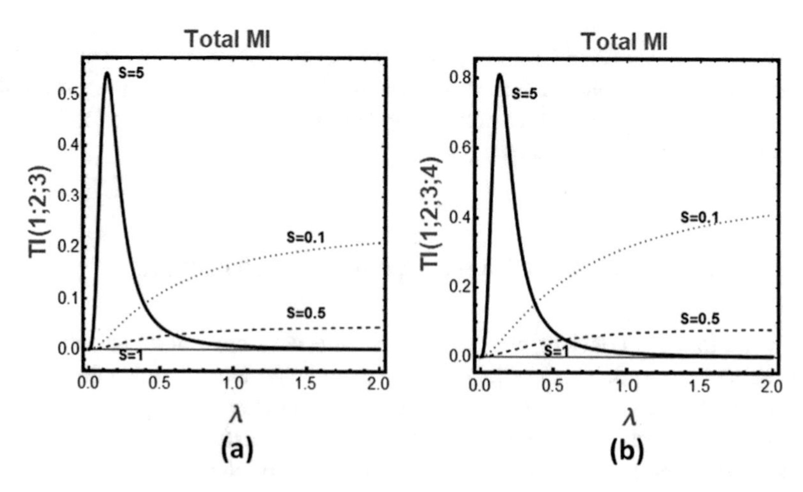

Fig. 5.69 The total triplet MI (**a**), and the quadruplet MI (**b**) as a function of λ, for different values of S

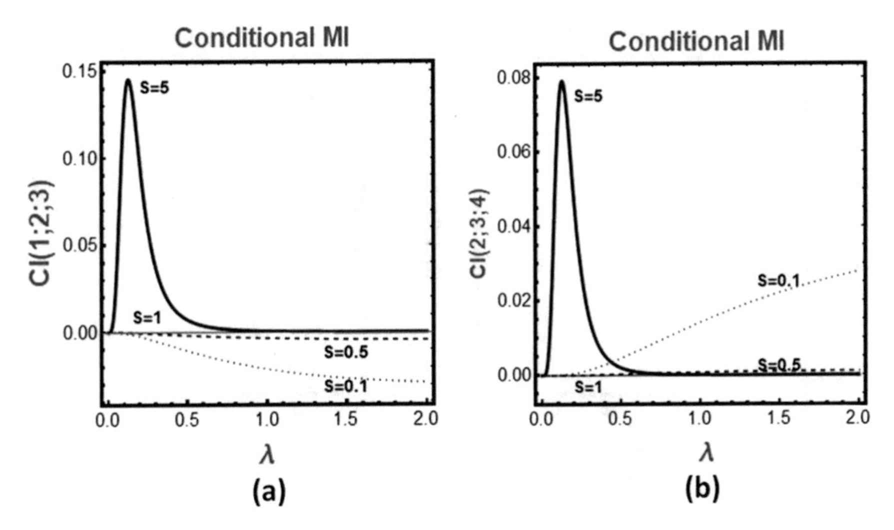

Fig. 5.70 The conditional triplet MI, CI(1;2;3) (**a**), and CI(2;3;4), MI (**b**) as a function of λ, for different values of S

and (1, 3)], but the latter has only two interactions [between (2, 3) and (3, 4)]. This seems to confirm the tentative conclusion regarding the negative sign of CI and the odd number of repulsive interaction.

If we accept this conclusion it would be difficult to explain the behavior of the quadruplet CI, Fig. 5.71. For small values of λ, it seems that $CI(2;3;4)$ is everywhere positive. However, for very large values of λ, we find that the $CI(2;3;4)$ becomes negative for $S = 0.1$. From these results we cannot associate the negative sign of CI with odd number of repulsion interactions.

5.5.5 *Tetrahedral Arrangement; Direct Interaction Between All Pairs of Sites*

In this and in the following sections we shall study the case of four sites arranged in different tetrahedral geometry. This means that all pairs of sites are at the same distance. When all the sites are occupied we have altogether six pair interactions.

All the probabilities, the various SMIs, and the pair-MI are quite similar to the case of a perfect square. Here, we show the binding isotherm for the tetrahedral case. Although all the curves are typical Langmuir isotherms, we present these in Fig. 5.72 for a later comparison of conformational changes in the subunits.

Figure 5.73 shows the triplet and the quadruplet total MI. As can be seen all the values are positive and the curves are quite similar to the case of perfect square.

Figure 5.74 shows the triplet and the quadruplet conditional MI. Here, there is a difference between the triplet and the quadruplet CI. For $S > 1$ all the values of

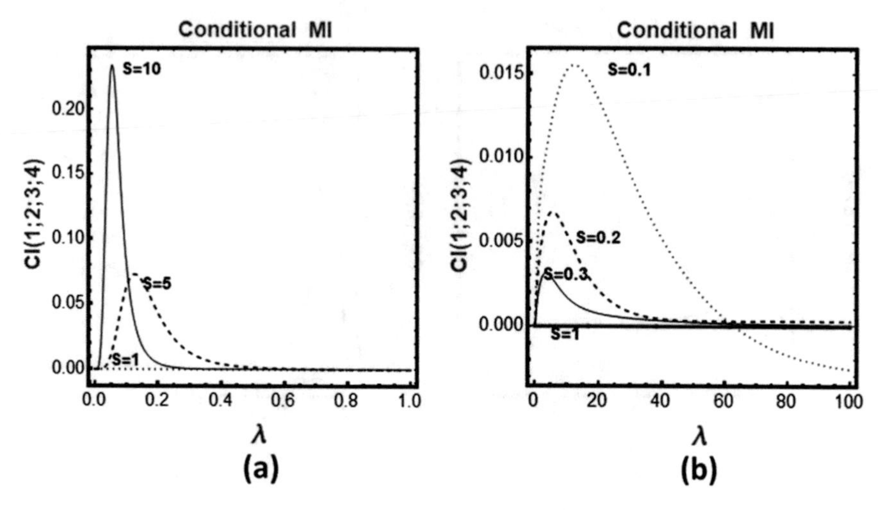

Fig. 5.71 The conditional quadruplet MI as a function of λ for different values of S

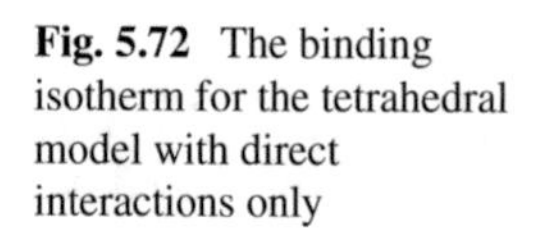

Fig. 5.72 The binding isotherm for the tetrahedral model with direct interactions only

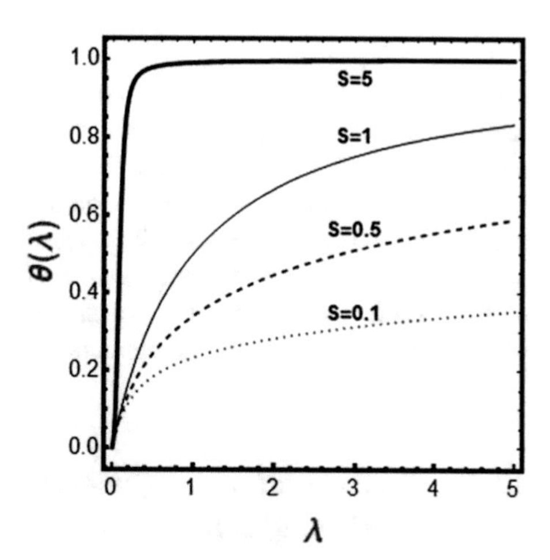

the CI are positive. For $S < 1$, we get negative values for the triplet CI, but positive values for the quadruplet CI.

We can conclude this section by declaring that in all cases the various probabilities, the SMIs, and the total MI (this includes all pair-MI) behave similarly as a function of λ. The only quantity which behaves differently when the interaction changes is the conditional MI (which is defined only for three or more sites). This quantity is sometimes negative, and sometimes positive. When the interaction is attractive (here, the factor S is larger than one), we always get a positive CI. When the interactions are repulsive ($S < 1$), we might get either positive or negative values of CI. Sometimes, the sign of the CI changes when λ changes. Therefore, there is no general conclusion

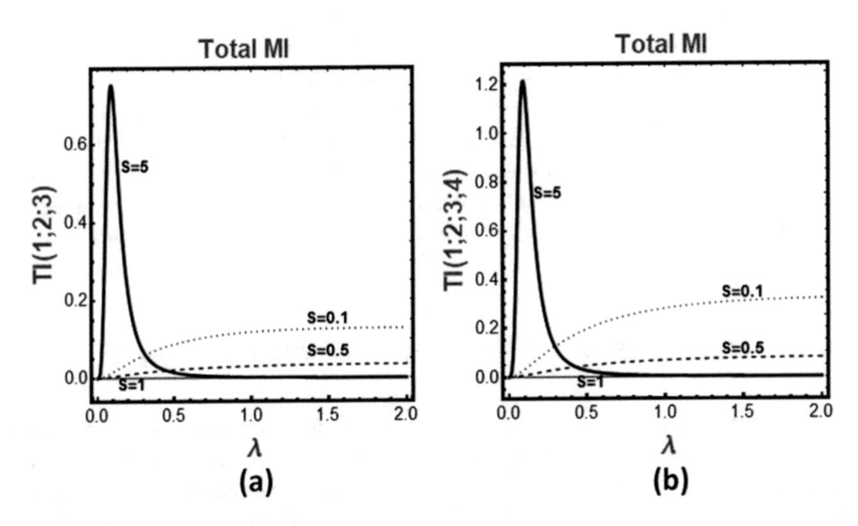

Fig. 5.73 The total triplet MI (**a**), and the quadruplet MI (**b**) for the tetrahedral model for different values of S

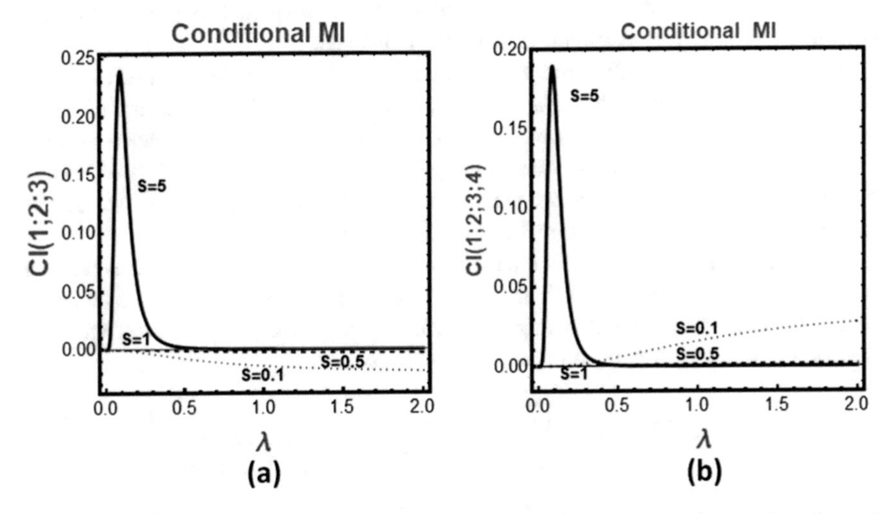

Fig. 5.74 The conditional triplet MI (**a**), and quadruplet MI (**b**) as a function of λ, for different values of S

that can be drawn between repulsive interaction and the sign of the CI. In addition, one can definitely say that there is no relationship between the sign of CI, and the extent of frustration in the system (see also Chap. 3).

As a final comment it should be said that in Chap. 3, we studied systems with *fixed* number of spins. In this chapter we also studied systems with *fixed number of sites*. However, for the same system with a fixed number of sites, the number of interacting ligands change with λ. For example, in the tetrahedral model there are always four binding sites, and six pair correlations. However, for small values of λ we have a mixture of molecules with different number of ligands, hence, also on average small number of interacting ligands. When λ increases, more and more of the sites are occupied, and the average number of interacting ligands also increase. Thus, when we try to associate the sign of CI with the number of attractive or repulsive interactions, one should keep in mind that these numbers are only average numbers; small, on average for small values of λ, and large, on average for larger values of λ.

5.6 Four-Site Systems with Indirect Interactions Only

In this section, we study systems with four binding sites, but the sites are distant from each other. This excludes the possibility of *direct* ligand-ligand interactions. Thus, all the correlations and MI in these systems are due to the indirect communication of information between the sites. The *information* communicated is simple; one site "knows" whether other sites are empty or occupied. The means of communication is through conformational changes in the subunits, induced by the binding of ligands.

As we have explained in the introduction of this chapter, this type of indirect communication is very common in biological systems.

In this system we shall compare the behavior of four arrangements of subunits; linear, square, and tetrahedral, Fig. 5.75. It should be noted that in the case of *direct* interactions between ligands, we have 3, 4, and 6 ligand-ligand interactions in the linear, square, and tetrahedral arrangement, respectively. In the case discussed in this chapter, we have again 3, 4, and 6 subunit-subunit *direct* interactions. However, the indirect interaction is always between all pairs of sites.

In the following subsections we will discuss only a few quantities, such as binding isotherms, SMI, and various MI. In each subsection we will compare the behavior of the system with different arrangements; linear, square and tetrahedral. We shall use the same parameters used in Sect. 5.3.3. These are:

$$h = \frac{q_H}{q_L}, \quad K = \frac{Q_H}{Q_L}, \quad \overline{K} = \frac{Q_{HH}}{Q_{LL}}, \quad \eta = \frac{Q_{LH}^2}{Q_{LL}Q_{HH}} \tag{5.77}$$

h measures the relative strength of binding of the ligands to sites L and H.

K measures the relative stability of the states L and H of the subunits.

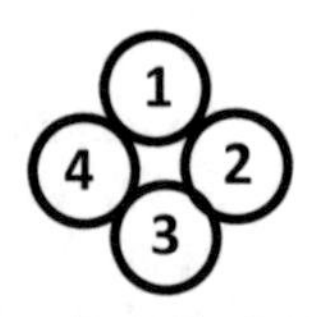

Fig. 5.75 The three different arrangements of the four subunits

$\overline{K}$ measures the relative strength of the subunit-subunit interactions for the pair of subunits in states LL and HH. η is the equilibrium constant for the reaction:

$$LL + HH \rightarrow 2LH$$

This parameter is important in understanding the efficiency of the transmission of information across the boundaries between the subunits. In addition, we have the absolute activity λ which is proportional to the density of the ligand in the phase which is at equilibrium with the adsorbed ligands.

Here, we have clearly five parameters that determine the behavior of the system. We shall use only a small sample of these parameters to highlight some of the difference between the various arrangements of the four sites.

5.6.1 *Binding Isotherms*

We start with a comparison of the binding isotherm for the three arrangements of the four subunits shown in Fig. 5.76. The binding isotherm is the most important quantity in biological systems. As we explained in the introduction to this chapter, the efficient transport of oxygen by Hb depends on the change in the value of θ between the two pressures in the loading and unloading terminals.

Suppose that the pressure at the lungs and at the cells are P_2 and P_1, respectively. Clearly, for this particular pair of pressures the most efficient transporting of the ligand will be for the tetrahedral arrangement. This is approximately the arrangement of the four subunits comprising the Hb molecule. It should be noted that in this molecule the actual distance between the oxygen molecules occupying two sites

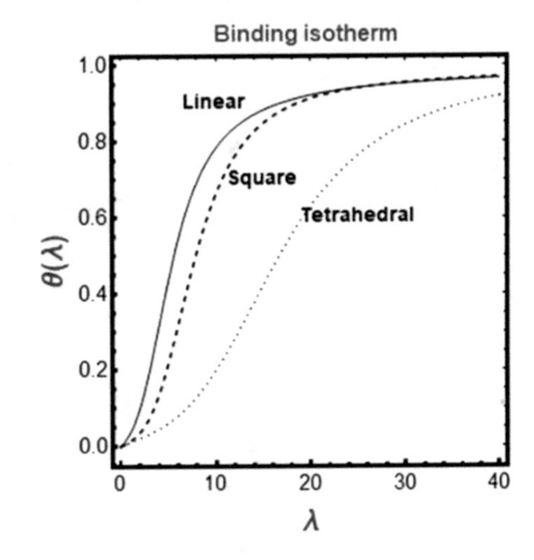

Fig. 5.76 The binding isotherm for the linear, square and tetrahedral models, as a function of the partial pressure of the oxygen

is such that direct interaction between the oxygen molecules is negligible. All the cooperativity in these systems is due to the indirect communication between the binding sites.

5.6.2 *The Various SMI in the Three Arrangements*

Before we discuss some specific SMIs, we note that all the probabilities in these systems behave as expected. For instance, the singlet probability increases monotonically with λ in all the systems. Note that there are two different singlet probabilities in the linear arrangement; site 1 and 4 are equivalent but are different from sites 2 and 3. For the square and tetrahedral arrangements all the singlet probabilities are the same.

There are also two different singlet SMI, for the edged sites (1 or 4) and the inner sites (2 or 3) in the linear arrangement. However, the general behavior of the SMI as a function of λ is the same for all the sites. Figure 5.77a shows a typical form of the dependence of singlet SMI as a function of λ. The curve starts at zero for $\lambda = 0$, and tends to 0 when $\lambda \to \infty$. This is a result of a single dominant state of occupancy at these two limits. As λ increases, the value of SMI rises sometimes, but not always to reach a maximum of about 1 (meaning that the two states of occupancy are equally probable), then decreases to 0 for large λ (at which the occupied site has probability 1). The pair, triplet, and quadruplet SMI have similar dependence on λ. The only difference is the maximum values which are nearly 2, 3 and 4, respectively. Figure 5.77b shows the quadruplet SMI for this system.

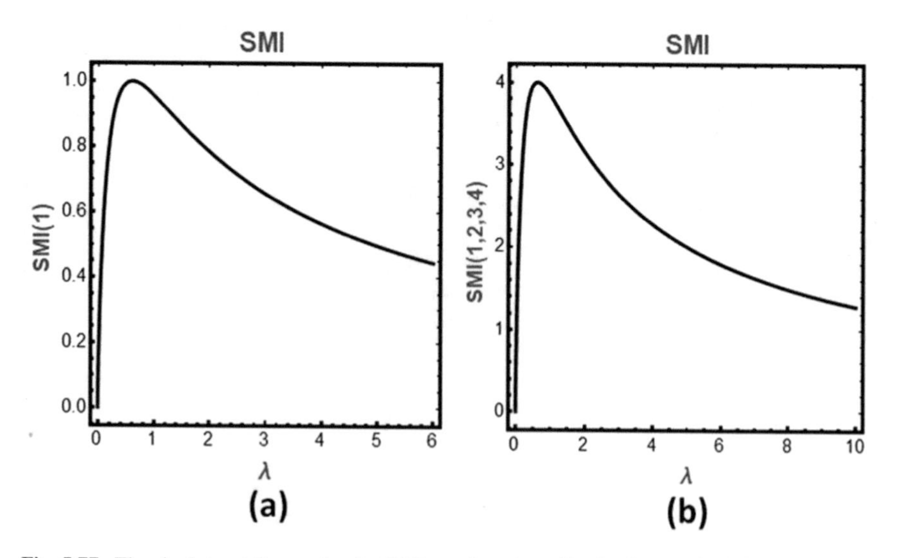

Fig. 5.77 The singlet and the quadruplet SMI as a function of λ, for the tetrahedral model

5.6.3 Total Mutual Information

Figure 5.78 shows a typical pair MI for the linear arrangement.

For this illustration we chose $h = 2$, $K = 1$, $\overline{K} = 2$, and various values of η. Note that when η increases from 0.1 to 0.3 to 1, the value of the MI decreases. At $\eta = 1$, there is no transmission of information across the boundaries between the subunits therefore $I(1; 2)$ is 0. When $\eta > 1$, the value of the MI starts to grow again. The λ dependence is the same for all the total MI, they all start at 0 for $\lambda = 0$, and tend to 0 at $\lambda \to \infty$. Remember that the total MI for any number of sites is by definition a positive number. They all measure the average logarithm of the corresponding correlation function. We next turn to the *conditional* MI where we find both positive and negative values of the CI, for both the triplet and the quadruplet sites. We shall show a few curves of both $CI(1; 2; 3)$, and $CI(1; 2; 3; 4)$ for the three arrangements of the four sites. The fact that we have both positive and negative values of CI, and the fact that the sign of the CI sometimes changes with λ, precludes any interpretation of the CI as a measure of either frustration or of the relative number of attractive or repulsive interactions. It is by no means clear at this stage whether a consistent interpretation of CI may be given which is valid for all systems.

Figure 5.79 shows some values of $CI(1; 2; 3)$, with parameters $\mathrm{K} = 0.1$, $\overline{K} = 5$, $\mathrm{h} = 2$, and $\eta = 0.1$, and $CI(1; 2; 3; 4)$, with parameters $\mathrm{K} = 1$, $\overline{K} = 2$, $\mathrm{h} = 2$, and $\eta = 0.01$, for the square arrangement. Figure 5.80 shows some values of $CI(1; 2; 3)$, with parameters $\mathrm{K} = 0.1$, $\overline{K} = 5$, $\mathrm{h} = 2$, and $\eta = 0.01$, and $CI(1; 2; 3; 4)$, with parameters $\mathrm{K} = 0.1$, $\overline{K} = 6$, $\mathrm{h} = 5$, and $\eta = 0.01$, for the square arrangement.

Figure 5.81 shows some values of $CI(1; 2; 3)$, with parameters $\mathrm{K} = 2$, $\overline{K} = 0.1$, $\mathrm{h} = 5$, and $\eta = 0.01$, and $CI(1; 2; 3; 4)$, with parameters $\mathrm{K} = 0.1$, $\overline{K} = 6$, $\mathrm{h} = 2$, and $\eta = 0.01$, for the tetrahedral arrangement.

The reader is urged to look at the figures, perhaps also compute some other values for different parameters, h, K, KK and η. Think about the meaning of each of the

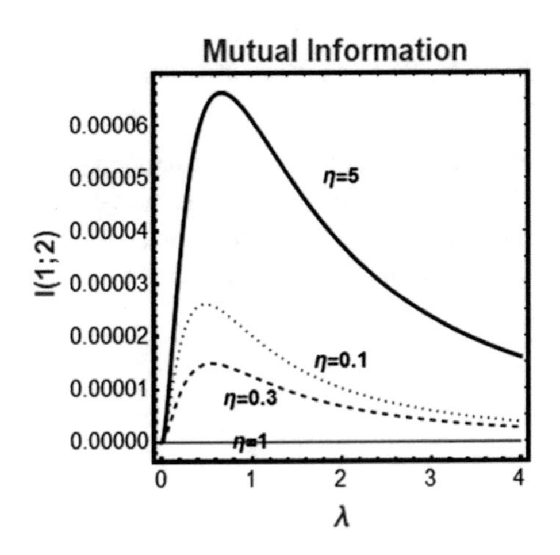

Fig. 5.78 The pair MI as a function of λ, for the tetrahedral model

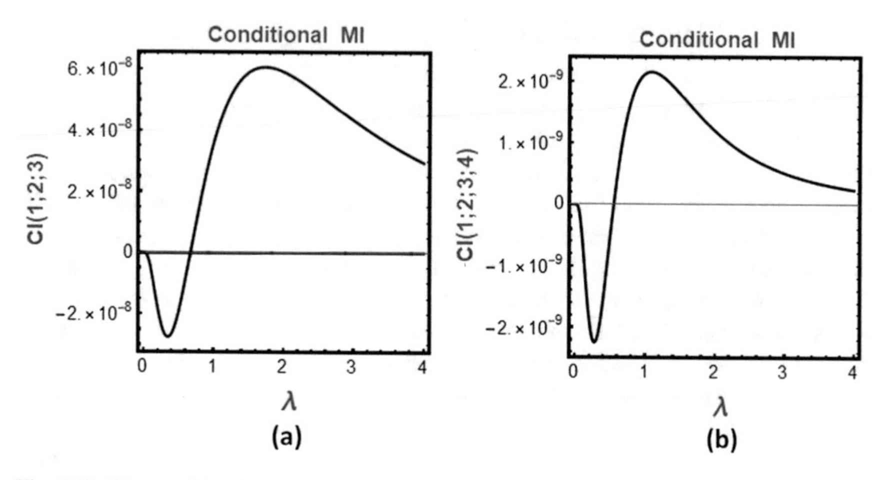

Fig. 5.79 The conditional triplet the quadruplet MI for the linear arrangement

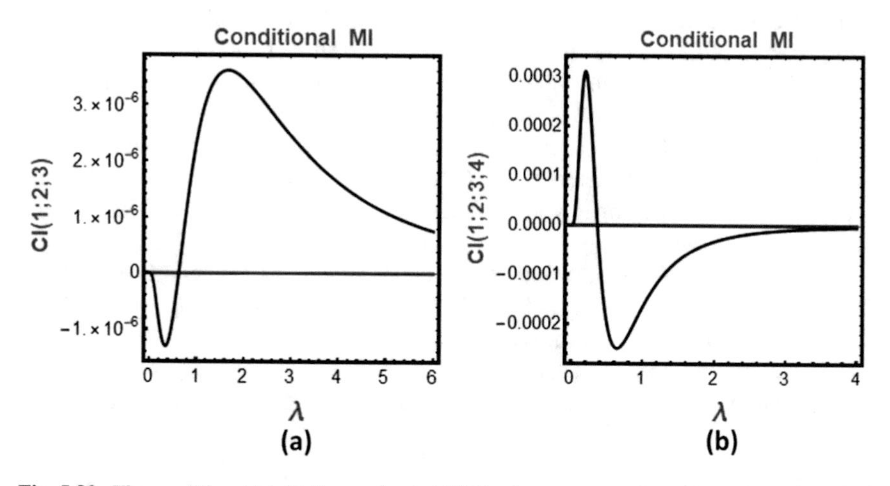

Fig. 5.80 The conditional triplet the quadruplet MI for the square arrangement

computed quantity and try to understand the behavior of these quantities when you change the parameters of the model. If you reach any insight, understanding, or interpretation of these values please write to me, and I will gladly share your views in a future publication of this book.

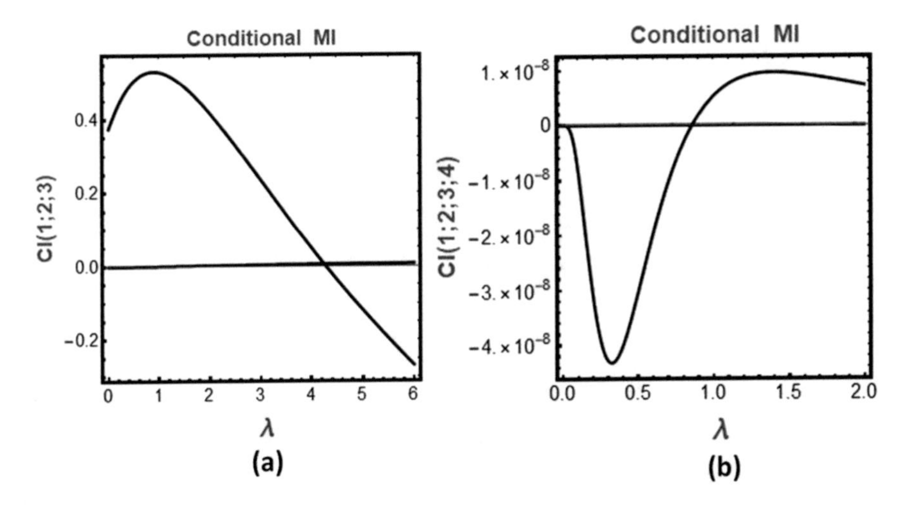

Fig. 5.81 The conditional triplet the quadruplet MI for the tetrahedral arrangement

References

1. Ben-Naim, A. (2001). *Cooperativity and regulation in biochemical systems*. Kluwer/Plenum Publications.
2. Adair, G. S. (1925). *Journal of Biological Chemistry, 63*, 529.
3. Pauling, L. (1935). The oxygen equilibrium of haemoglobin and its structural interpretation. *Proceedings of the National academy of Sciences of the United States of America, 21*, 186.
4. Antonini, E., & Brunori, M. (1971). *Haemoglobin and myoglobin in their reaction with ligands*. North-Holland.
5. Imai, K. (1982). *Allosteric effects in haemoglobin*. Cambridge University Press.
6. Monod, J., Changeux, J. P., & Jacob, F. (1963). *Journal of Molecular Biology, 6*, 306.
7. Monod, J., Wyman, J., & Changeux, J. P. (1965). *Journal of Molecular Biology, 12*, 88.
8. Ben-Naim, A. (1992). *Statistical thermodynamics for chemists and biochemists*. Plenum Press.
9. Ben-Naim, A. (2014). *Statistical thermodynamics, with applications to life sciences*. World Scientific.
10. Ben-Naim, A. (2017). *Information theory, part I: An introduction to the fundamental concept*. World Scientific.
11. Matsuda, H. (2000). *Physical Review E, 62*, 3096.

Chapter 6
Calculations of the "Best-Guess" Probability Distribution Using Shannon's Measure of Information

Following the publication of Shannon's article on "A Mathematical Theory of Communication" in 1948, [1, 2], Jaynes [3, 4], developed the so-called "*maximum-entropy principle in statistical mechanics.*" As I have argued in Ben-Naim [5], this principle should be called Maximum-Shannon-Measure-of-Information, which we abbreviate as MaxSMI, instead of calling it the Maximum-Entropy. The thermodynamic entropy has nothing to do with the idea underlying the method of MaxSMI. Unlike the previous chapters where we use known probability distributions to estimate the SMI, here we have a kind of "inverse" problem; to estimate the probability distribution by the method of MaxSMI.

In this chapter we present a few examples of the application of the MaxSMI principle. This is essentially a method of *guessing* the "*best*" or the *least biased* distribution function, using whatever information is available on the distribution. In 1967, Katz wrote [6]:

> Information theory approach is not a miracle device to arrive at a solution of any statistical problem. It is only a general way to formulate the problem itself consistently. It most often happens that the "Best-guess" coincided with the educated guess of those who practice guessing as an art.

This "best-guess," according to Katz, is the one that is consistent with the "*truth, the whole truth and nothing but the truth.*" In my view, a better way of describing what the "best-guess" means, is to say that it is consistent with the *knowledge, the whole knowledge and nothing but the knowledge* we have on the system under study. This is closer to Jaynes' original description of the merits of the maximum entropy principle.

We shall start with the simplest example of an unknown distribution; the distribution of two outcomes of tossing a coin. In fact, the solution of this example is almost trivial. Next, we shall discuss the case of a die, on the distribution of which we know a few moments. Finally, we shall apply the method of MaxSMI to a real experimental system; the distribution of micelles-size. We shall see how we can get a "better" or improved guess of the distribution using experimentally available information on the system.

A. Ben-Naim, *Information Theory and Selected Applications*,
https://doi.org/10.1007/978-3-031-21276-5_6

6.1 The Probability Distribution of an Unfair Coin

Suppose we have a coin, on each of its sides is a number. For simplicity, assume that the two numbers are: "2" and "3."

Note that if the two sides are distinguished by different letters (say, H and T), or by different colors (say, blue and red), then there is no meaning to the *average* result, or to any moments of the distribution [7, 8].

If we do not have any information on the distribution, then the MaxSMI procedure is to maximize the SMI:

$$H = -\sum p_i \log p_i \tag{6.1}$$

Subject to the normalization condition:

$$\sum p_i = 1 \tag{6.2}$$

This procedure leads to the result that the best-guess of the distribution is:

$$p_i = p_2 = 0.5 \tag{6.3}$$

Next, suppose we are told that someone threw the coin (with the same two sides but with unknown probabilities) numerous times and obtained the average result of 2.3:

$$\langle i \rangle = \sum i p_i = 2p_1 + 3p_2 = 2.3 \tag{6.4}$$

In this particular example, the *average* given in (6.4) *determines* the distribution. The reason is that in this particular case we have only one unknown, say p_1, and the second probability is $p_2 = 1 - p_1$. Therefore, Eq. (6.4) is an equation for one unknown (p_1), which we can rewrite as:

$$2p_1 + 3(1 - p_1) = 2.3 \tag{6.5}$$

The solution of this equation is:

$$p_1 = 0.7,\ p_2 = 0.3 \tag{6.6}$$

Thus, in this case the given average in (6.4), uniquely determines the probability distribution in (6.6).

It is easy to show that the same result is obtained by the method of MaxSMI. In the following we shall use the method of Lagrange multipliers, and assume for simplicity the natural logarithm.

We define the auxiliary function:

$$F = -\sum p_i \log p_i + \lambda_1 \left[1 - \sum p_i\right] + \lambda_2 \left[2.3 - \sum i p_i\right] \tag{6.7}$$

The condition of maximum F is:

$$\frac{\delta F}{\delta p_i} = -\log p_i - 1 - \lambda_1 - \lambda_2 i = 0 \tag{6.8}$$

Which, for our case is equivalent to:

$$p_1 = e^{-1-\lambda_1-\lambda_2} \tag{6.9}$$

Since $p_2 = 1 - p_1$ we can substitute in Eq. (6.5) to obtain the same result as in (6.6). For more details on this procedure see the next section.

6.2 A Six-Face Die with Three Different Numbers of Dots; 1, 2 and 3

Consider a regular die where two of its faces has a dot each, the other two faces have two dots on each face, and the remaining faces have three dots each. We denote by p_1, p_2 and p_3 the probabilities of finding one, two, and three dots, respectively. The SMI for this die is:

$$H = -\sum_{i=1}^{3} p_i \log p_i \tag{6.10}$$

with the normalization condition

$$\sum_{i=1}^{3} p_i = 1 \tag{6.11}$$

If we do not have any other information on this die, we know that the resulting distribution in this case is the uniform distribution. This is obtained by maximizing the function:

$$F = -\sum p_i \log p_i + \lambda_1 \left(1 - \sum p_i\right) \tag{6.12}$$

The condition for the maximum is:

$$\left(\frac{\partial F}{\partial p_i}\right) = -\log p_i - 1 - \lambda_1 = 0 \tag{6.13}$$

or equivalently:

$$p_i = \exp(-1 - \lambda_1) \tag{6.14}$$

Note that in the sums on the right hand side of (6.12), i is a running index, but when we take the derivative in (6.13) the derivative is for each specific p_i. Note that in the result (6.14), p_i is independent of the index i. Hence, it is a constant. Substituting in Eq. (6.11) we obtain:

$$\sum p_i = 3\exp[-1 - \lambda_1] = 1 \tag{6.15}$$

Hence,

$$p_i = \frac{1}{3} \text{ for } i = 1, 2, 3 \tag{6.16}$$

Next, we assume that we know the average result is 2.5, i.e., we know that:

$$\langle i \rangle = \sum_{i=1}^{3} i p_i = 2.5 \tag{6.17}$$

Note that if the die is fair, i.e. the distribution is uniform, Eq. (6.16) then the average result will be $\frac{1}{3}(1 + 2 + 3) = 2$. In Eq. (6.17) we are given an average larger than 2. Therefore, we can guess that the distribution is such that the faces with larger number of dots have higher probability.

Clearly, having only the average (6.17) and the normalization condition (6.11) does not determine uniquely the distribution. We have two Eqs. (6.15 and 6.17) for three variables. Therefore, in order to calculate the "best-guess" distribution, we apply the MaxSMI method. We define the function:

$$F = -\sum p_i \log p_i + \lambda_1\left[1 - \sum p_i\right] + \lambda_2\left[2.5 - \sum i p_i\right] \tag{6.18}$$

The condition of maximum is:

$$\frac{\partial F}{\partial p_i} = -\log p_i - 1 - \lambda_1 - i\lambda_2 = 0 \tag{6.19}$$

or equivalently:

$$p_i = \exp[-1 - \lambda_1 - i\lambda_2] \tag{6.20}$$

The two coefficients λ_1 and λ_2 can be determined by substituting (6.20) into Eqs. (6.11) and (6.17), i.e.

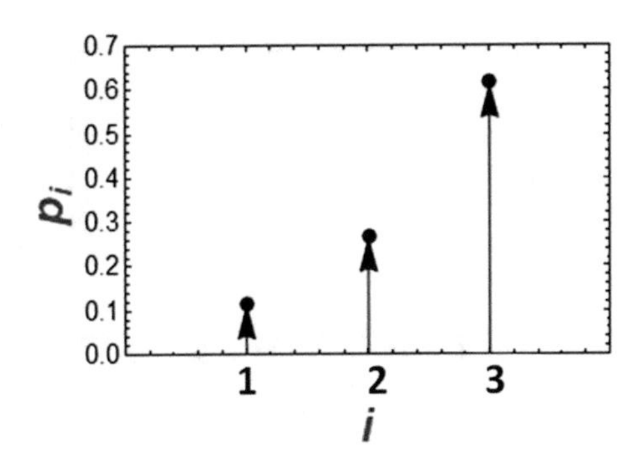

Fig. 6.1 The probability distribution for the case discussed in Sect. 6.2

$$1 = \sum p_i = \exp[-1 - \lambda_1] \sum \exp[-\lambda_2 i] \tag{6.21}$$

$$2.5 = \sum i p_i = \exp[-1 - \lambda_1] \sum i \exp[-\lambda_2 i] \tag{6.22}$$

There are two equations for λ_1 and λ_2, or equivalently:

$$1 = y \sum x^i \tag{6.23}$$

$$2.5 = y \sum i x^i \tag{6.24}$$

where $y = \exp[-1 - \lambda_1]$ and $x = \exp[-\lambda_2]$. Equations (6.23) and (6.24) are two equations for the two variables x and y. The solution for the two unknown is: $x \simeq 2.3$, $y \simeq 0.0505$. The corresponding probabilities are:

$$p_1 = 0.116,\ p_2 = 0.267,\ p_3 = 0.61 \tag{6.25}$$

These are shown in Fig. 6.1.

As expected, there is higher probability for the outcomes 3 compared to the outcomes 1 or 2.

6.3 Probability Distribution of an Unfair Regular Six-Face Die

The next example is a regular six-face die. We want to find the best probability distribution for this case. What is the best-guess probability distribution, given only the normalization condition? Clearly, there are infinite number of possible. Examples:

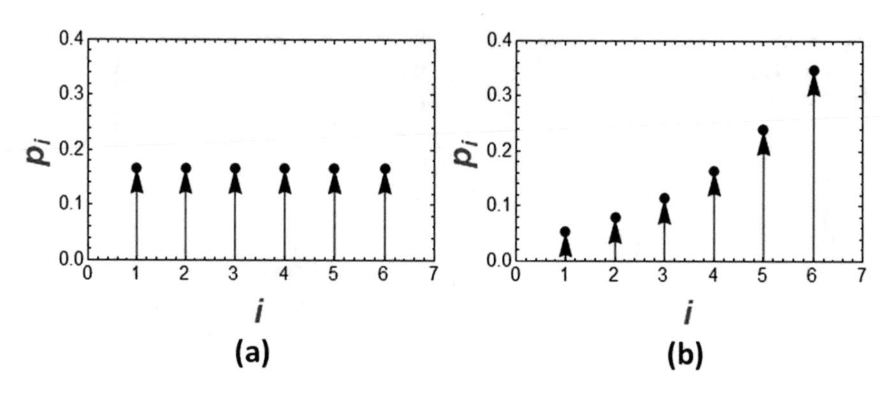

Fig. 6.2 The probability distributions for the case discussed in Sect. 6.3

$$\begin{array}{l} 1,\ 0,\ 0,\ 0,\ 0,\ 0 \\ \frac{1}{2},\ \frac{1}{2},\ 0,\ 0,\ 0,\ 0 \\ \frac{1}{4},\ \frac{1}{4},\ \frac{1}{4},\ \frac{1}{4},\ 0,\ 0 \\ \frac{1}{6},\ \frac{1}{6},\ \frac{1}{6},\ \frac{1}{6},\ \frac{1}{6},\ \frac{1}{6}, \end{array} \tag{6.26}$$

Obviously, the first example is highly biased. Why? Because nothing in the available information tells us that the first outcome is the certain event. Similarly, the second example is somewhat less biased, but again there is nothing in the available information that tells us that the first two outcomes are the only possible events. Similarly, the third choice is less biased than the second, but again there is nothing in the available information which tells us that the last two outcomes are impossible. Thus, what is left in this list is the fourth choice (of course there are more possibilities which we did not list). This is also the least biased choice. It is easy to show that this distribution is obtained by maximizing the SMI subject to the only normalization condition. This distribution is shown in Fig. 6.2a.

Next, suppose we have an additional information. We are told that measurements have been done and it was found that the average outcome was 4.5. In this case, the distribution must satisfy the following two conditions:

$$\sum_{i=1}^{6} p_i = 1 \tag{6.27}$$

$$\sum_{i=1}^{6} ip_i = 4.5 \tag{6.28}$$

In this case we have only two equations and six unknown quantities (which is the probability distribution). There are many solutions to this problem. The "best-guess" distribution, is obtained by maximizing the SMI, subject to the two conditions (6.27) and (6.28). In this case we can write two equations:

$$1 = y \sum x^i \tag{6.29}$$

$$4.5 = y \sum i x^i \tag{6.30}$$

These are two equations with the two unknowns: x and y. One can solve these equations to obtain the "best-guess" distribution. This distribution is shown in Fig. 6.2b.

6.4 Calculation of an Approximate Micelle-Size Distribution Using Experimental Data

In this section, I will discuss very briefly an application of the (SMI), to calculate a distribution of a more complicated system; the micelle-size-distribution (MSD). The method consists of analyzing the experimental data on the dependence of the osmotic pressure of a surfactant, on the total concentration of the surfactant in the solution. We first show that the moments of the MSD are related to the derivatives of these functions. Then, by using a limited number of such moments, and by using the mathematical procedure of maximizing the SMI, we compute the "best-guess" of the MSD.

It is well-known that surfactant molecules such as long fatty acids aggregate in aqueous solutions. At some high concentration they form micelles. There have been a few attempts to compute the micelle size distribution (MSD) by using a statistical mechanical approach. These calculations were all based on some very simplifying assumptions concerning the nature of the solute–solute and solvent–solvent interactions. With these assumptions, supplemented by some further mathematical approximations, one can compute the MSD as well as some thermodynamic quantities, such as the osmotic pressure of the surfactant solution. Since there is no experimental way of measuring the MSD, the only way of assessing the success of a particular model is the comparison between the computed and the experimental values of the thermodynamic quantities. Such a comparison is not a stringent test of either the physical model or the computational procedure. Many different models may lead to the same thermodynamic quantities. In this section, we show how we can obtain approximate information on the MSD from knowledge of experimental data on the osmotic pressure of a surfactant solution. The extent of accuracy and reliability of the MSD we obtained by this method would depend on the details and accuracy of the experimental data.

In what follows we outline a procedure for calculating what may be referred to as the "best-guess" MSD. In particular, we assume that we have data on the osmotic pressure, but in principle other thermodynamic data, such as activity coefficients, vapor pressure, etc. could be used as well. The method consists of analyzing the experimental data in essentially three steps:

First, we compute of the monomer concentration. Second, we compute the *moments* of the MSD. And finally we estimate the "best-guess" MSD which is consistent with the available experimental data. In the latter, we use the idea of maximizing the SMI defined on the MSD.

Throughout the following analysis, we use the assumption that the system, water-surfactant, is an *associated-ideal* solution. This means that all interactions among the aggregates are neglected. This assumption appears to be valid for dilute solutions of non-ionic surfactants, but may not be valid for high concentrations, or for ionic surfactant solutions, where non-ideality effects should be taken into account. The two fundamental equations for a system of a surfactant in solvent are:

$$\rho_T = \sum_{i=1}^{\infty} i\rho_i \tag{6.31}$$

$$\beta\pi = \rho_C = \sum_{i=1}^{\infty} \rho_i \tag{6.32}$$

where ρ_i is the number (or molar) density of the aggregates consisting of i monomers, ρ_T is the total concentration of the surfactant in the solution. π is the osmotic pressure and $\beta = (kT)^{-1}$, with k is the Boltzmann constant (or the gas constant, if we use molarities concentrations throughout). In this section we can assume that $k = 1$. ρ_C is the total density of the aggregates in the system, sometimes referred to as the apparent molarity of the surfactant. Using the assumption of ideality of the solution, we may relate each of the ρ_i to the monomer concentration ρ_1. To do so, we assume a reaction of the form:

$$i \text{ monomers} \rightleftharpoons \text{aggregates of size } i. \tag{6.33}$$

And the corresponding equilibrium constant:

$$\rho_i = K_i \rho_1^i \tag{6.34}$$

where K_i is the *equilibrium constant*, for the association reaction (6.33). In terms of ρ_1, Eqs. (6.31) and (6.22) may be rewritten as

$$\rho_T = \sum_{i=1} i K_i \rho_1^i \tag{6.35}$$

$$\rho_C = \sum_{i=1} K_i \rho_1^i \tag{6.36}$$

Let k_n be the equilibrium constant for the reaction in which we add one monomer to an aggregate of size n.

$$A_{n-1} + M \rightarrow A_n \tag{6.37}$$

$$k_n = \frac{\rho_n}{\rho_1 \rho_{n-1}} \tag{6.38}$$

Here, A_n is an aggregate consists of n monomers. In the simplest association process, the addition of one monomer to an aggregate involves the same amount of work, independently of n, i.e. we take

$$k_n = k_2 \text{ for all } n \tag{6.39}$$

The general relation between K_n and k_i is:

$$K_n = \prod_{i=2}^{n} k_i \tag{6.40}$$

This equation reduces to $K_n = (k_2)^{n-1}$ for the simplest model in Eq. (6.39). In principle, k_2 may be obtained from experimental data. Having a plot of ρ_1 as a function of ρ_T (see below) one finds:

$$\lim_{\rho_T \to 0} \left(\frac{\partial^2 \rho_1}{\partial {\rho_T}^2} \right) = -4k_2 \tag{6.41}$$

In the most general case we define the correction function G_n as:

$$K_n = k_2^{n-1} G_n \tag{6.42}$$

Assuming that k_2 is known from experimental sources we may rewrite Eqs. (6.35) and (6.36) in a reduced form, namely

$$y = \sum_i i G_i x^i \tag{6.43}$$

$$f = \sum G_i x^i \tag{6.44}$$

where we defined: $x = k_2 \rho_1$, $y = k_2 \rho_T$ and $f = \beta \pi k_2$.

6.5 Computation of the Monomer Concentration

We assume that experimental data on the osmotic pressure as a function of the total concentration of the surfactant are available, i.e. we have the values of ρ_C as a function of ρ_T. Both ρ_T and ρ_C are related through in Eqs. (6.43) and (6.44). From Eqs. (6.35) and (6.36) we can easily obtain:

$$\frac{\partial \rho_C}{\partial \rho_T} = \frac{\partial \ln \rho_1}{\partial \ln \rho_T} \tag{6.45}$$

Integrating this identity between $\rho_T(1)$ and $\rho_T(2)$, we get:

$$\ln \rho_1(2) - \ln \rho_1(1) = \int_{\rho_T(1)}^{\rho_T(2)} \frac{\partial \rho_C}{\partial \rho_T} d\ln \rho_T \tag{6.46}$$

We now choose $\rho_T(1)$ to be very small, so that at this concentration we have $\rho_T(1) \approx \rho_1(1)$. With this assumption we may rewrite Eq. (6.46) as:

$$\begin{aligned} \ln \rho_1(2) &= \ln \rho_T(1) + \ln \rho_T(2) - \ln \rho_T(2) + \int_{\rho_T(1)}^{\rho_T(2)} \left(\frac{\partial \rho_C}{\partial \rho_T} \right) d\ln \rho_T \\ &= \ln \rho_T(2) + \int_{\rho_T(1)}^{\rho_T(2)} \left[\frac{\partial \rho_C}{\partial \rho_T} - 1 \right] d\ln \rho_T \end{aligned} \tag{6.47}$$

Since the integrand is zero when $\rho_T \to 0$, we may replace the lower limit of the integral by $\rho_T(1) = 0$, to obtain the final result; for any concentration ρ_T we write:

$$\rho_1 = \rho_T \exp\left\{ \int_0^{\rho_T} \left[\frac{1}{\rho_T} \frac{\partial \rho_C}{\partial \rho_T} - \frac{1}{\rho_T} \right] d\rho_T \right\} \tag{6.48}$$

Clearly, the entire right-hand-side of Eq. (6.48) may be computed from experimental data, hence ρ_1 may be calculated for each ρ_T.

6.6 Calculation of the Moments of the Micelle-Size-Distribution

The micelle-size-distribution MSD is defined by:

$$P_i = \frac{\rho_i}{\sum_{j=1}^{\infty} \rho_j} \tag{6.49}$$

The l'th moment of the distribution is defined by:

$$M_l = \sum_{i=1}^{\infty} i^l P_i = \frac{\sum_{i=1} i^l K_i \rho_1^i}{\sum_{i=1} \rho_i} = \langle i \rangle^l \tag{6.50}$$

It is convenient to define the sum:

$$N_l = \sum_{i=1} i^l \rho_i = \sum_{i=1} i^l K_i \rho_1^i \tag{6.51}$$

Which gives the l'th moment of the distribution:

$$\langle i \rangle^l = \frac{N_l}{N_0} \tag{6.52}$$

In particular, we have:

$$\langle i \rangle^0 = 1 \tag{6.53}$$

$$\langle i \rangle^1 = \frac{N_1}{N_0} = \frac{\rho_T}{\rho_C} \tag{6.54}$$

Next, we take the derivative:

$$\frac{\partial \rho_C}{\partial \rho_T} = \frac{\partial \rho_C}{\partial \rho_1}\frac{\partial \rho_1}{\partial \rho_T} = \frac{\sum_i i K_i \rho_1^{i-1}}{\sum_i i^2 K_i \rho_1^{i-1}} = \frac{N_1}{N_2} = \frac{\langle i \rangle}{\langle i \rangle^2} \tag{6.55}$$

Having computed $\langle i \rangle$ from Eq. (6.54), we obtain $\langle i^2 \rangle$ from knowledge of the first derivative of ρ_C as a function of ρ_T. This procedure may be continued to higher derivatives, for instance, the second derivative gives:

$$\frac{\partial^2 \rho_C}{\partial \rho_T^2} = \frac{\langle i^2 \rangle^2 - \langle i \rangle \langle i^3 \rangle}{\rho_C \langle i^2 \rangle^3} \tag{6.56}$$

Once we have computed $\langle i \rangle$ and $\langle i^2 \rangle$, we can calculate $\langle i^3 \rangle$ from Eq. (6.56). This procedure becomes more complicated, and less useful as we proceed to higher moments. A simpler procedure for a numerical computation is to use the sum N_l defined in Eq. (6.51). Thus using the derivative:

$$\frac{\partial \rho_C}{\partial \rho_T} = \frac{N_1}{N_2} \tag{6.57}$$

From which we compute N_2 (since $N_1 = \rho_T$). Having N_2 as a function of ρ_T we take its first derivative:

$$\frac{\partial N_2}{\partial \rho_T} = \frac{N_3}{N_2} \tag{6.58}$$

from which N_3 may be determined at each ρ_T. Next, we take

$$\frac{\partial N_3}{\partial \rho_T} = \frac{N_4}{N_2} \tag{6.59}$$

to obtain N_4 at each ρ_T, and so on. In general, having computed N_k at the k-th step, we may proceed to obtain N_{k+1}, at the k + ith step from

$$\frac{\partial N_k}{\partial \rho_T} = \frac{N_{k+1}}{N_2} \tag{6.60}$$

wherefrom N_{k+1} may be computed. Clearly, at each step the moments may be computed from the sums N_k, through

$$M_k = \frac{N_k}{N_0} \tag{6.61}$$

The procedure outlined above shows that if we have a very accurate and dense set of measurements, then we could obtain as many moments as we wish. In practice however, the procedure may be useful to obtain only the first few moments. This allows one to estimate the best-guess of the MSD which is consistent with the available experimental information.

6.7 Computation of the "Best-Guess" MSD

We assume that we have already analyzed the experimental curves of ρ_C as a function of ρ_T, and obtained L moments of the distribution, say

$$M_l = \sum_{i=1}^{\infty} i^l P_i l = 0, 1, 2, 3, \cdots, L \tag{6.62}$$

Clearly, If we know the distribution, we can easily compute all its moments. The converse is also true, if we have *all* the moments of a distribution, then we can reconstruct the distribution itself. Here however, we have only the first few moments, and therefore we can expect to get an approximate distribution. What is the "best-guess" distribution that we can obtain given a few moments? This kind of question may be answered by the well-known procedures of maximum SMI (known also as the Max Entropy method). We first define the SMI of any distribution by:

$$H = -\sum_{i=1}^{n} P_i \ln P_i \tag{6.63}$$

In writing Eq. (6.63) we assumed that the aggregate of maximum size consists of n monomers. The "best-guess" of a distribution is obtained by maximizing H in (6.63) subject to the constraint that L moments are known. The mathematical problem is thus to find an extremum of the auxiliary function:

$$F = H + \sum \lambda_l M_l \tag{6.64}$$

where the parameters,λ_l are the Lagrange multipliers. The condition for extremum is:

$$\frac{\partial}{\partial P_j}\left[-\sum P_i \ln P_i + \sum_{l=0}^{L} \lambda_l \sum i^l P_i\right] = 0, \text{ for each } j = 1, 2, \cdots n. \tag{6.65}$$

Performing the differentiation with respect to P_j we obtain the condition

$$P_j = \exp[\lambda_0 - 1]\exp\left[\sum_{l=0}^{L} \lambda_l j^l\right] \tag{6.66}$$

Thus, if we know all the λ_l's we may compute the required distribution P_j. The λ_l's may be determined from the $L + 1$ equations defining the moments, namely:

$$M_k = \sum_{j=1}^{n} j^k P_j = \sum_{j=1} j^k \exp[\lambda_0 - 1]\exp\left[\sum_{l=1}^{L} \lambda_l j^l\right] \tag{6.67}$$

for $k = 0, i, \cdots L$. These are $L + 1$ equations for the $L + 1$ unknowns.

We can use the fact that $M_0 = 1$ to eliminate λ_0, i.e.

$$1 = \exp[\lambda_0 - 1]\sum_{j=1}^{n} \exp\left[\sum_{j=1}^{n} \lambda_l j^l\right] \tag{6.68}$$

Dividing each M_k in Eq. (6.67) by Eq. (6.68) we obtain

$$M_k = \frac{\sum_{j=1}^{n} j^k \exp\left[\sum_{l=1} \lambda_l j^l\right]}{\sum_{j=1}^{n} \exp\left[\sum_{l=1} \lambda_l j^l\right]} \tag{6.69}$$

Now we have L equations with the unknowns $\lambda_1 \cdots \lambda_L$. Once these are determined, one can use Eq. (6.68) to compute λ_0, and thereby all of the P_j's in Eq. (6.66) can be calculated.

Some Numerical Examples

The illustrations presented in the following use a theoretical model as an input, i.e. we start from a choice of a series of K_i. Then we choose values of ρ_T at close intervals. Next, we invert Eq. (6.35) to obtain the corresponding values of ρ_1 as a function of ρ_T. Once ρ_1 is known we compute all the ρ_i's through Eq. (6.34), and hence the input-distribution ρ_i in Eq. (6.49). Also, we can use Eq. (6.36) to compute ρ_C at each ρ_T, and obtain the (theoretical) curve of ρ_C as a function of ρ_T for this particular model.

We now treat the curve of $\rho_C = \rho_C(\rho_T)$ as if it were a (hypothetical) experimental curve. We use it to follow the procedure as outlined in the previous sections. First, we compute ρ_1 as a function of ρ_T from Eq. (6.48) and compare the results with the values computed from the model. Second, we compute the moments of the distribution by the method outlined above, and again compare with the moments computed from the model. Finally, we use this procedure to compute the "best-guess" of a distribution, and compare it with our input distribution. This procedure has one clear-cut advantage over a straightforward application of the method on *real* experimental data. In the latter, whenever we obtain a distribution, there is no way of assessing its relevance to the real distribution of the experimental solution. On the other hand, by using a model-input-distribution, and following the above outlined procedure, we obtain an output distribution which may be compared with the input distribution. In this way the effectiveness of the procedure is tested, and a detailed examination of its dependence on various parameters (number of points, their density, accuracy, etc., may be studied).

In the following we use dimensionless quantities, hence we can use either Eqs. (6.43) and (6.44), or the pair of Eqs. (6.35) and (6.36) (with $k_2 = 1$). The simplest model to start with it the *monodispersed* case, i.e.

$$K_i = \begin{cases} 1 & for\; i = 1 \\ 1 & for\; i = n_{\max} \\ 10^{-10} & for\; 1 < i < n_{\max} \end{cases} \tag{6.70}$$

Here n_{max} is the largest aggregate which is assumed to exist in the solution. In our first "experiment" we take $n_{\max} = 10$. We use 50 points for ρ_T. At each point we compute ρ_1 from Eq. (6.35) and substitute into Eq. (6.36) to compute ρ_C. The two curves obtained are shown in Fig. 6.3a, and similarly in Fig. 6.3b for $n_{\max} = 20$.

Now we use the values of ρ_C as a function of ρ_T to follow the procedure of converting these data into MSD. First, we compute ρ_1 from Eq. (6.48). we find that the results obtained agree with the original values, and are almost indistinguishable from the results plotted in Fig. 6.3. Next, we computed the moments $M_0 \cdots M_5$ (note,$M_0 = 1, M_1 = \rho_T/\rho_C$). We note that the agreement between the computed moments and the original ones is satisfactory. There seems to be a better agreement for the lower moments, but even for M_5 the agreement is within less than one percent of their values. Finally, we proceed to compute the MSD. Here, we start with a low value of ρ_T where we could guess initial values of λ_i's. then use these as input parameter in an iterative procedure to computeλ_i's which solve the set of Eqs. (6.69). When increasing the total concentration, we use the values of theλ_i's from a previous concentration as an initial guess, for the next higher concentration. The initial value ofλ_i's for the first ρ_T are chosen as follows. We rewrite the distribution as:

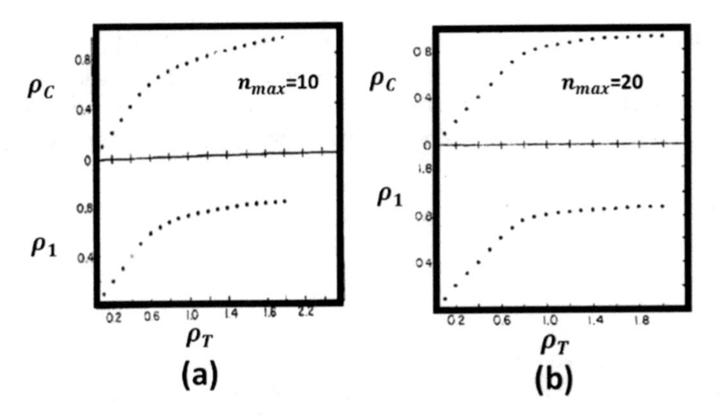

Fig. 6.3 Thedensitiesρ_C and ρ_1 as a function of the total ρ_T for the monodispersed case with **a** $n_{\max} = 10$, and **b** for $n_{\max} = 20$

$$P_j = \frac{\rho_j}{\sum \rho_i} = \frac{K_j \rho_1^j}{\rho_C} \tag{6.71}$$

$$\ln P_j = j \ln \rho_1 - \ln \rho_C + \ln K_j \tag{6.72}$$

Clearly, in the limit of $\rho_T \to 0$, both $\rho_1 \sim \rho_T$, and $\rho_C \sim \rho_T$, i.e., only monomers exist in the solution. Thus, for $j = 1$ we must have $P_j = 1$, and all other P_j's $= 0$. For $\rho_T \sim 0$ we must still assume that the first two terms on the right-hand-side of Eq. (6.72) will be the leading terms, i.e.

$$\ln P_j = j \ln \rho_1 - \ln \rho_1 \rho_T \sim 0) \tag{6.73}$$

Comparing with Eq. (6.66), we get:

$$\ln P_j = (\lambda_0 - 1) + \sum_{l=0}^{L} \lambda_l j^l \tag{6.74}$$

We arrive at the initial guess for the parameters:

$$\begin{aligned} \lambda_0 &= 1 - \ln \rho_1 \\ \lambda_1 &= \ln \rho_1 \\ \lambda_2 &= \lambda_3 = \cdots \lambda_2 = 0 \end{aligned} \tag{6.75}$$

These parameters are used in the iterative procedure for computation of λ_i's at the first ρ_T, and so forth. Because of the inaccuracy of the procedure we expect that the MSD as determined at each concentration will be very inaccurate. Therefore, we used an averaging procedure to determine K_i (since ρ_1 and ρ_C are known, the knowledge of either P_i or K_i are equivalent). The advantage of this procedure is that whereas the MSD is dependent on the concentration ρ_T, the series of values of

K_i are independent of the concentration of the surfactant (which follows from the ideality assumption adopted at the beginning of our treatment). The details of the calculations may be found in Ben-Naim [9]. Here we only note that by using this procedure we get very good agreement between the input, and the output quantities K_i.

6.8 Conclusion

In this chapter we saw several examples of how one can get the "best-guess" of a probability distribution based on limited information on the moments of the distribution.

References

1. Shannon, C. E. (1948). A mathematical theory of communication. *Bell System Technical Journal, 27*, 379.
2. Shannon, C. E., & Weaver, W. (1949). *The mathematical theory of communication.* The University of Illinois Press.
3. Jaynes, E. T. (1957). *Physical Review, 106*, 620.
4. Jaynes, E. T. (1957). *Physical Review, 108*, 171.
5. Ben-Naim, A. (2017). *Information theory, part I: An introduction to the fundamental concept.* World Scientific.
6. Katz, A. (1967). *Principles of statistical mechanics: The informational theory approach.* W. H. Freeman.
7. Ben-Naim, A. (2010), *Discover entropy and the second law of thermodynamics. A playful way of discovering a law of nature.* World Scientific.
8. Ben-Naim, A. (2015). *Discover probability. How to use it, how to avoid misusing it, and how it affects every aspect of your life.* World Scientific.
9. Ben-Naim, A. (1982). Computation of the micelle-size-distribution from experimental measurements. In B. Lindman & K. L. Mittal (Eds.), *Proceedings of the International Symposium on Surfactants in Solutions, Lund, Sweden*, Plenum Press, New York.

Index

A. Ben-Naim, *Information Theory and Selected Applications*,
https://doi.org/10.1007/978-3-031-21276-5

Printed in the United States
by Baker & Taylor Publisher Services